编委会

主 编

俞汉青

编 委

（以姓氏拼音排序）

陈洁洁	刘 畅	刘武军
刘贤伟	卢 姝	吕振婷
裴丹妮	盛国平	孙 敏
汪雯岚	王楚亚	王龙飞
王维康	王允坤	徐 娟
俞汉青	虞盛松	院士杰
翟林峰	张爱勇	张 锋

"十四五"国家重点出版物出版规划重大工程

生物质废弃物污染控制与清洁转化

Pollution Control and Clean Conversion of the Biomass Wastes

刘武军 著

中国科学技术大学出版社

内 容 简 介

本书全面介绍了生物质热解过程中污染元素的热解行为及机制、污染物的产生及释放调控机制,重点分析了含污染物的生物质在热解过程中主要污染元素(重金属、氮、磷)的分布、转化和迁移过程及其机制,考察了重金属对生物质热解的催化促进作用,并探讨了将氮、磷元素富集于生物炭相中,进而制备氮、磷、重金属原位掺杂多孔碳材料等高价值产品的调控方法。

图书在版编目(CIP)数据

生物质废弃物污染控制与清洁转化/刘武军著.—合肥:中国科学技术大学出版社,2022.3

(污染控制理论与应用前沿丛书/俞汉青主编)

国家出版基金项目

"十四五"国家重点出版物出版规划重大工程

ISBN 978-7-312-05395-5

Ⅰ.生⋯　Ⅱ.刘⋯　Ⅲ.生物质—废物处理—研究　Ⅳ.X7

中国版本图书馆 CIP 数据核字(2022)第 031457 号

生物质废弃物污染控制与清洁转化
SHENGWUZHI FEIQIWU WURAN KONGZHI YU QINGJIE ZHUANHUA

出版	中国科学技术大学出版社 安徽省合肥市金寨路96号,230026 http://www.press.ustc.edu.cn https://zgkxjsdxcbs.tmall.com
印刷	安徽联众印刷有限公司
发行	中国科学技术大学出版社
开本	787 mm×1092 mm　1/16
印张	17.75
字数	398 千
版次	2022 年 3 月第 1 版
印次	2022 年 3 月第 1 次印刷
定价	110.00 元

总　序

建设生态文明是关系人民福祉、关乎民族未来的长远大计,在党的十八大以来被提升到突出的战略地位。2017年10月,党的十九大报告明确提出"污染防治"是生态文明建设的重要战略部署,是我国决胜全面建成小康社会的三大攻坚战之一。2018年,国务院政府工作报告进一步强调要打好"污染防治攻坚战",确保生态环境质量总体改善。这都显示出党和国家推动我国生态环境保护水平同全面建成小康社会目标相适应的决心。

当前,我国环境污染状况有所缓解,但总体形势仍然严峻,已严重制约了我国经济社会的持续健康发展。发展以资源回收利用为导向的污染控制新理论与新技术,是进一步推动污染物高效、低成本、稳定去除的发展方向,已成为国家重大战略需求和国际重要学术前沿。

为了配合国家对生态文明建设、"污染防治攻坚战"的一系列重大布局,抢占污染控制领域国际学术前沿制高点,加快传播与普及生态环境污染控制的前沿科学研究成果,促进相关领域人才培养,推动科技进步及成果转化,我们组织一批来自多个"双一流"大学、活跃在我国环境科学与工程前沿领域、有影响力的科学家共同撰写"污染控制理论与应用前沿丛书"。

本丛书是作者团队承担的国家重大重点科研项目(国家重大科技专项、国家863计划、国家自然科学基金)和获得的重大科技成果奖励(2014年国家自然科学奖二等奖、2020年国家科学技术进步奖二等奖)的系统总结,是作者团队攻读博士学位期间取得的重要的前沿学术成果(全国百篇优秀博士论文、中国科学院优秀博士论文等)的系统凝练,是一套系统反映污染控制基础科学理论与前沿高新技术研究成果的系列图书。本丛书围绕我国环境领域的污染物生化控制、转化机理、无害化处置、资源回收利用等亟须解决的一些重大科学问题与技术问题,将物理学、化学、生物学、材料学等学科的最新理论成果以及前沿高新技术应用到污染控制过程中,总结了我国目前在污染控制领域（特别是废水和固废领域）的重要研究进展，探索、建立并发展了常温空气阴极燃料电池、纳米材料、新兴生物电化学系统、新型膜生物反应器、水体污染物的化学及生物转化，以及固体废弃物污染控制与清洁转化等方面的前沿理论与技术，形成了具有广阔应用前景的新理论和新方法，为污染控制与治理提供了理论基础和科学依据。

"污染控制理论与应用前沿丛书"是服务国家重大战略需求、推动生态文明建设、打赢"污染防治攻坚战"的一套丛书。其出版将有利于促进最前沿的科研成果得到及时的传播和应用，有利于促进污染治理人才和高水平创新团队的培养，有利于推动我国环境污染控制和治理相关领域的发展和国际竞争力的提升；同时为环境污染控制与治理实践提供新思路、新技术、新材料，也可以为政府环境决策、强化环境管理、履行国际环境公约等提供科学依据和技术支撑，在保障生态环境安全、实施生态文明建设、打赢"污染防治攻坚战"中起到不可替代的作用。

编委会

2021 年 10 月

前　言

中国每年生产大约15亿吨的废弃生物质（农作物秸秆、木屑、杂草）。这些废弃生物质蕴藏的能源相当于10亿吨的标准煤，此外其中所含的氮、磷、钾元素的肥力值高于中国每年生产化肥的总肥力值，是一个巨大的资源宝库。但是，由于高的劳动成本、低的经济效益以及日益严格的环境法规，目前废弃生物质的回收利用并没有得到广泛的推广。相反，大部分的生物质都是直接废弃，部分通过焚烧获取能量，不仅能量利用效率低，而且造成了严重的环境污染。具体表现在：① 生物质的不完全燃烧将释放大量的持久性有机污染物（氧化多环芳烃、二噁英等）和颗粒污染物（PM 2.5，PM 10），是造成中国近期大雾霾的主要原因之一；② 生物质燃烧产生的烟雾降低空气的能见度，从而影响道路和空中交通安全，甚至造成严重的交通事故。因此，迫切需要研发出环境友好、经济可行的方法实现这些大量存在的废弃生物质的回收利用。快速热解是一项极具前景的生物质资源化成熟技术，如果将其应用于对污染生物质的处置及资源回收，将会具有重要的环境和经济意义。

本书针对生物质废弃物热化学处置和资源回收基础研究中的科学问题，以热解过程解析和热解产物优化为中心，开展了系列研究，主要包括以下内容：

（1）解析了生物质废弃物在热解过程中主要杂元素（重金属、氮、磷）的分布、转化和迁移过程及其机制，并探索出抑制热解污染物的产生，将氮、磷、重金属元素富集于生物炭相中，进而制备氮、磷、重金属原位掺杂多孔碳材料的调控手段。

（2）针对生物油品质较差、利用价值低的缺点，发展了零价金属原位还原、催化有机物、电解等环境友好可持续的转化体系，实现了生物油品质的优化提升，并将生物油中大量存在的小分子转化为高价值化学品，实现其清洁增值转化。

（3）以生物炭为平台材料，通过前体调控手段，设计制备了一系列生物炭基功能材料,并成功应用于污染物催化转化、二氧化碳捕集、清洁能源转化与储存等领域。

由于编者水平有限,书中难免会有疏漏和不尽如人意之处,敬请相关专家和读者批评指正。

编　者

2021年10月

目 录

总序 —— i

前言 —— iii

第 1 章
绪论 —— 001

1.1 概述 —— 003

1.2 生物质的基本组成及其热解机理 —— 005

1.3 生物质废弃物热化学转化过程及其机理 —— 006

1.4 生物质热解产物的品质优化提升 —— 011

1.5 生物质热解过程中污染物的产生、迁移、转化过程与机制 —— 022

第 2 章
生物质热解过程中污染物的迁移转化过程及其机理解析 —— 051

2.1 生物质表面化学修饰强化污染物去除及其机理研究 —— 053

2.2 铅污染生物质热解铅元素迁移转化过程及机理 —— 069

2.3 湿地植物生物质热解氮磷营养元素迁移转化过程及其机制解析 —— 077

2.4 生物质和废弃电子垃圾塑料共热解过程中溴化阻燃剂的分解转化机理探索 —— 090

第 3 章
生物质热解液体产物的品质提升及其机理研究 —— 123

3.1 电催化 5-羟甲基糠醛的阳极氧化制备 2,5-呋喃二甲酸 —— 125

3.2 常温常压下零价金属对生物油的品质提升及其作用

机制研究 —— 133

3.3 快速热解铜负载生物质选择性提高生物油的品质
　　—— 145

第 4 章
生物炭基功能材料的设计合成 —— 171

4.1 $MgCl_2$ 负载生物质热解制备介孔碳支撑的纳米
　　氧化镁材料及其对 CO_2 捕集性能研究 —— 173

4.2 生物质热解制备碳基磁性固体酸及催化性能评价
　　—— 186

4.3 快速热解碳化法从富氮湿地植物生物质制备
　　氮掺杂多孔碳材料及其电化学储能性能研究
　　—— 198

4.4 泡沫生物炭电极的制备及其电催化氧化 5-HMF
　　—— 213

4.5 铁氮共掺杂生物炭材料设计合成及其电催化性能
　　—— 228

第 1 章

绪论

1.1 概述

中国每年生产大约15亿吨的废弃生物质(农作物秸秆、木屑、杂草)。[1]这些废弃生物质蕴藏的能源相当于10亿吨的标准煤,此外其中所含的氮、磷、钾元素的肥力值高于中国每年生产化肥的总肥力值,是一个巨大的资源宝库[2]。但是,由于高的劳动成本、低的经济效益以及日益严格的环境法规,目前废弃生物质的回收利用并没有得到广泛的推广。相反,大部分的生物质都是直接废弃,部分通过焚烧获取能量,不仅能量利用效率低,而且造成了严重的环境污染。具体表现在:① 生物质的不完全燃烧将释放大量的持久性有机污染物(氧化多环芳烃、二噁英等)[3-5]和颗粒污染物(PM 2.5,PM 10)[6-7],是造成中国近期大雾霾的主要原因之一[8];② 生物质燃烧产生的烟雾降低空气的能见度,从而影响道路和空中交通安全,甚至造成严重的交通事故[9]。因此,迫切需要研发出环境友好、经济可行的方法,实现这些大量存在的废弃生物质的回收利用。

目前主要使用的生物质资源化方法包括微生物发酵转化和热化学转化两种。[10]其中微生物发酵转化包括有氧发酵和厌氧消化,广泛应用于将废弃生物质转化为高附加值化学品(乙醇、丁醇、有机酸等)[11-13]。然而,微生物发酵转化过程通常涉及一系列的代谢反应,反应过程耗时较长且不容易控制。此外,发酵过程微生物代谢释放出大量的二氧化碳和甲烷等温室气体[14],将大大加速全球气候变暖的趋势。相对而言,由于操作简单、易于控制,热化学转化方法在生物质资源化利用方面具有更好的应用前景[15-16]。多种热化学转化方法,包括热解[17-24]、气化[25-31]、水热液化[32-35]等,已经广泛应用于将废弃生物质转化为燃料或化学品。在这些方法中,热解因能耗低(快速热解过程中的能耗大约只有生物质所含能量的10%)[36]且易于工业化大规模生产,是目前应用最为广泛的生物质热化学转化方法[37]。大量生物质通过热解转化而得的生物油(bio-oil),通过精制之后可以用作石油等化石燃料的替代品[38],同时也可以用作提取高附加值化学品(酚类、糠醛等)的原料[39]。热解剩余的生物炭(biochar)可以作为原料制备功能碳材料[40-42],也可以直接用作燃料或土壤修复剂[43],具有很高的环境和经济价值。

热解是指在无氧环境、适宜温度下(一般为400~700 ℃),将生物质热分解从而获得生物油和生物炭的过程[44-45]。热解产物的性质和产率受到诸多因素的影响,包括热解温度、升温速率和反应时间等[44]。热解过程根据升温速率和反应时间的不同,可以分为慢速热解、快速热解以及闪速热解[46-47]。慢速热解,通常是在较慢的升温速率($5\sim7\,K\cdot min^{-1}$)和较长的停留时间(大于 1 h)下对生物质进行碳化,这个过程获得的主要产物为生物炭,其产率大大高于生物油和热解气。而快速热解过程的升温速率大大增加($300\sim500\,K\cdot min^{-1}$),停留时间大为降低($0.5\sim10\,s$),获得的液体产物生物油的量显著增加[44,48]。由于生物油可以用作燃料或化学品原料,因此快速热解的应用前景更为广泛。为了尽可能多地获得生物油,快速热解的基本要求[44]如下:① 由于生物质的热导率通常很低,为了保证热解过程中生物质表面和内部的热传导过程顺利进行,通常需要尺寸很小的生物质颗粒(<3 mm)才可以满足要求;② 热解温度需要较为精确地控制在 $600\sim900\,K$,从而使得液体产物的量最大化;③ 为了保证热解蒸气成分不至于进一步裂解成小分子气体,热解蒸气停留时间需要控制在 10 s 以内;④ 热解蒸气需要快速冷凝,从而获得生物油。闪速热解可以认为是一种改进的快速热解过程,过程中热解温度通常在 $900\sim1200\,K$,升温速率高达 $1000\,K\cdot min^{-1}$。为了适应如此高的升温速率,生物质进料的颗粒必须非常细,通常小于 0.5 mm。由于超高的热解温度和升温速率,闪速热解的主要产物是热解气体,包括一些小分子烃类(如 CH_4,C_2H_4,C_2H_6 等)、H_2 和 CO 等。这三种热解形式的基本参数如表 1.1 所示。

表 1.1 不同热解形式的基本参数归纳

	慢速热解	快速热解	闪速热解
升温速率($K\cdot min^{-1}$)	5~7	300~800	>1000
热解温度(K)	500~1200	600~900	600~1200
进料尺寸(mm)	5~50	<3	<0.5
进料停留时间	>1 h	0.5~10 s	<2 s
常用反应器	固定床	流化床或固定床	流化床
主要产物	生物炭	生物油	热解气

1.2 生物质的基本组成及其热解机理

生物质的组成主要包括三个基本的结构单元：纤维素、半纤维素和木质素[49]。图1.1(a)所示是这三个结构单元在生物质中的分布[38-39,50]。纤维素是一类由β-D-吡喃葡萄糖(β-D-glucopyranose)单元通过β-糖苷键连接而成的均聚物[51-52]，是生物质中存在最多的物质，占生物质总量的40%～50%[53]。半纤维素是一类无定形的支链聚合物，其聚合度相对纤维素而言大为降低，并且结构单元也更为复杂。半纤维素占生物质总量的15%～30%[54]，其结构单元包括多种五碳糖(β-D-木糖和α-L-树胶醛糖)、六碳糖(β-D-甘露糖、β-D-葡萄糖和α-D-半乳糖)和糖醛酸(α-D-葡萄醛、α-D-4-O-甲基葡萄醛和α-D-半乳糖醛酸)。占生物质总量15%～30%的木质素是另一类无定形的聚合物，其结构根据生物质种

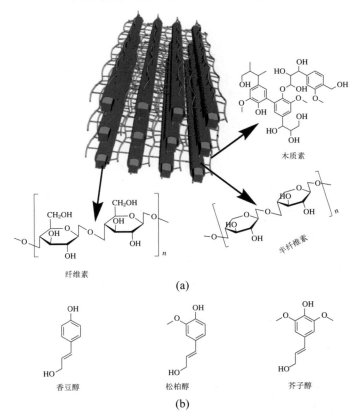

图 1.1 (a) 生物质中纤维素、半纤维素和木质素的分布情况及其结构示意图；(b) 木质素的单体的化学结构

类的不同而变化多样。木质素的结构单元主要是一类具有苯丙烷结构的芳香醇类，包括香豆醇（p-coumaryl alcohol）、松柏醇（coniferyl alcohol）和芥子醇（sinapyl alcohol）。尽管木质素占生物质总量不足30%，但是由于其碳含量较高，其所含能量占生物质所含总能量的40%以上。除了以上三种基本的构造单元之外，生物质中还含有少量的其他物质，包括蛋白质、叶绿素和一些无机元素等。

1.3 生物质废弃物热化学转化过程及其机理

1.3.1 生物质热解机理

生物质的三种基本结构单元的热解过程机理和分解途径主要受到热解反应器的类型、热解温度和升温速率的影响[55-62]。一般来说，在一个典型的热解过程中，半纤维素首先分解，其分解温度为470~530 K，其后是纤维素开始分解，分解温度为510~620 K，最后随着热解温度升高到550 K以上，木质素也开始分解[46]。对于半纤维素，其热解过程开始于聚合物链的断裂，形成一系列的低聚糖，随后这些低聚糖的糖苷键进一步裂开，得到的解聚分子经过重排，形成1,4-脱水吡喃木糖（1,4-anhydro-D-xylopyranose）[63]。这是半纤维素热解过程中的一个重要中间体，可以进一步分解产生很多小分子的热解产物，包括呋喃类和一些二碳、三碳的产物等（图1.2）[64-68]。和半纤维素类似，纤维素的热解过程也是开始于聚合物链的断裂，形成大量聚合度较低的葡萄糖低聚物，随后这些低聚物的糖苷键进一步热分解，并且经历分子间的重排，形成左旋葡聚糖中间体[69]，然后再经过一系列的反应，形成最终热解产物（图1.3）[70-78]。

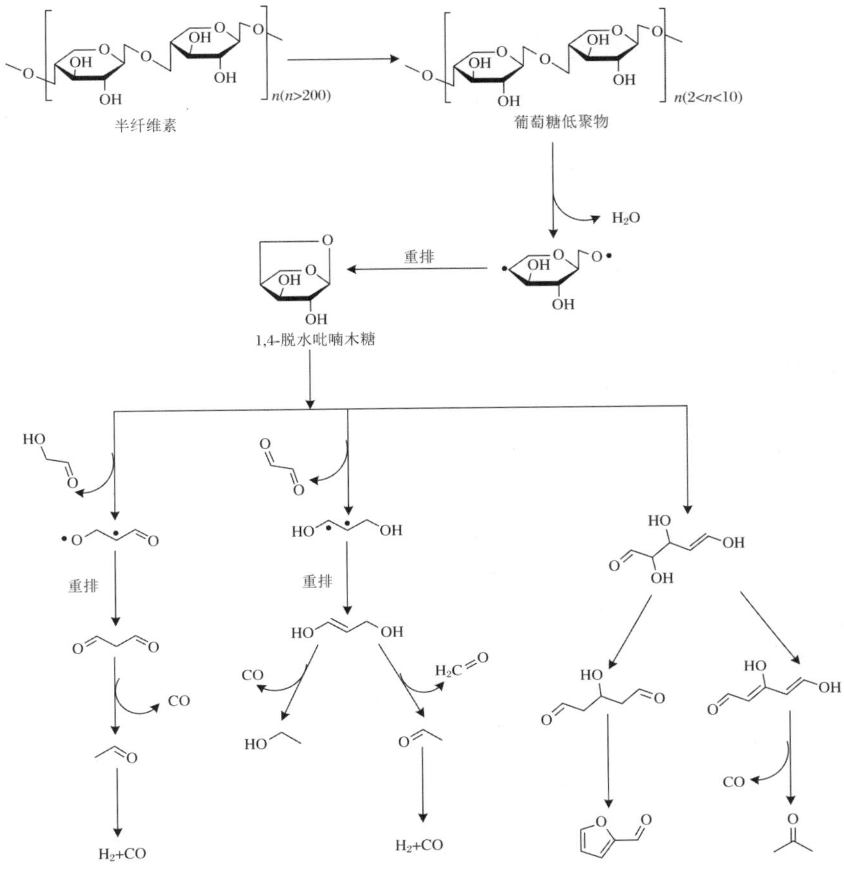

图 1.2　半纤维素热解过程机理示意图

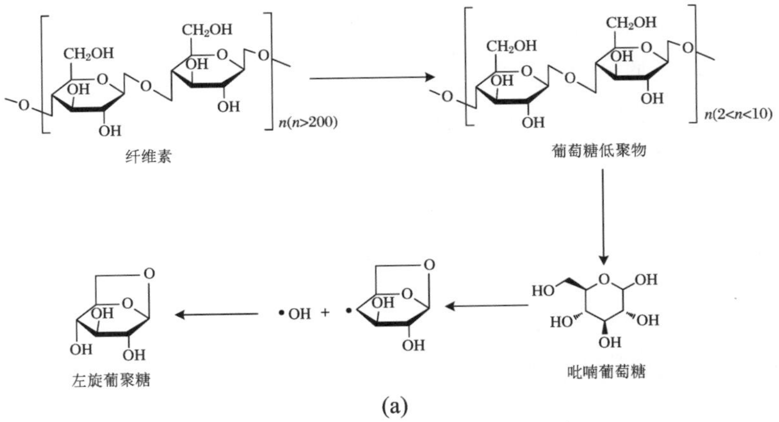

图 1.3　热解过程中纤维素分解至左旋葡聚糖的示意图(a)及热解过程中左旋葡聚糖可能的分解途径示意图(b)

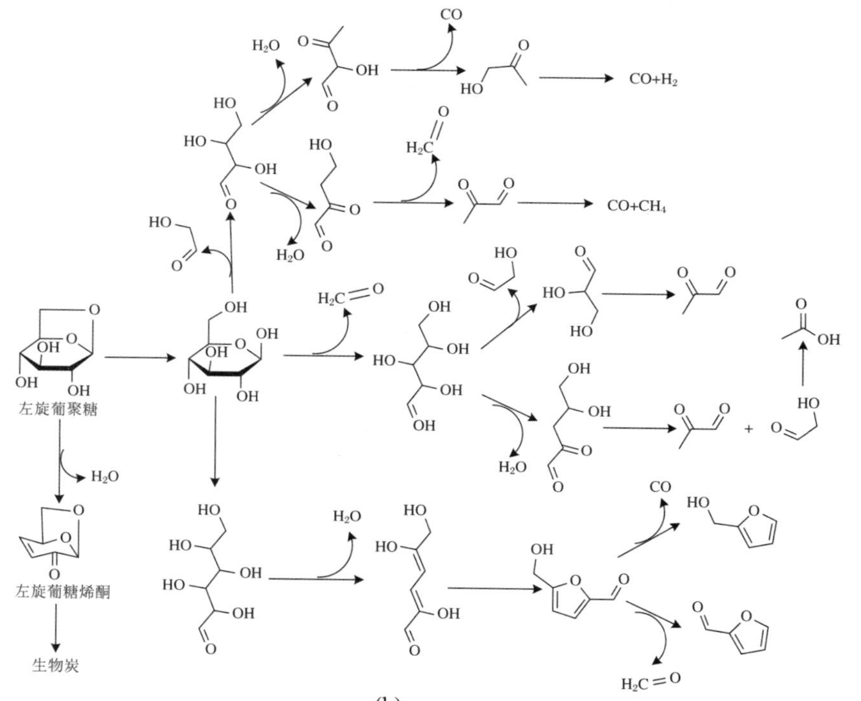

(b)

续图 1.3　热解过程中纤维素分解至左旋葡聚糖的示意图(a)及热解过程中左旋葡聚糖可能的分解途径示意图(b)

与纤维素和半纤维素相比,木质素的结构复杂很多,因此其热解机理也将更为复杂。目前比较公认的木质素热解机理主要是自由基反应主导的分解途径[79-82]。如图 1.4 所示,木质素分子中 β-O-4 连接断裂产生的自由基是产生自由基链式反应的开始,随后这些自由基可以从其他含有弱的 C—H 或者 O—H 键的结构中捕获质子,从而形成热解产物(例如香草醛和一些酚类化合物)[80,83],然后自由基重新转移到其他结构使得链式反应能够不断进行,当两个自由基相互结合形成稳定化合物时链式反应才能终止。

1.3.2
生物质热解过程的影响因素

温度是影响生物质热解过程的主要因素之一[46,84-85]。温度最基本的作用就是为生物质分解提供足够的热能,在低的温度下(<300 ℃),生物质的热解过程主要发生在其结构中的杂原子(N 和 O)位点,产生大量的重质焦油和无定形碳,而在高的温度下(>550 ℃),生物质的所有组分都将会彻底发生分解,产生大量

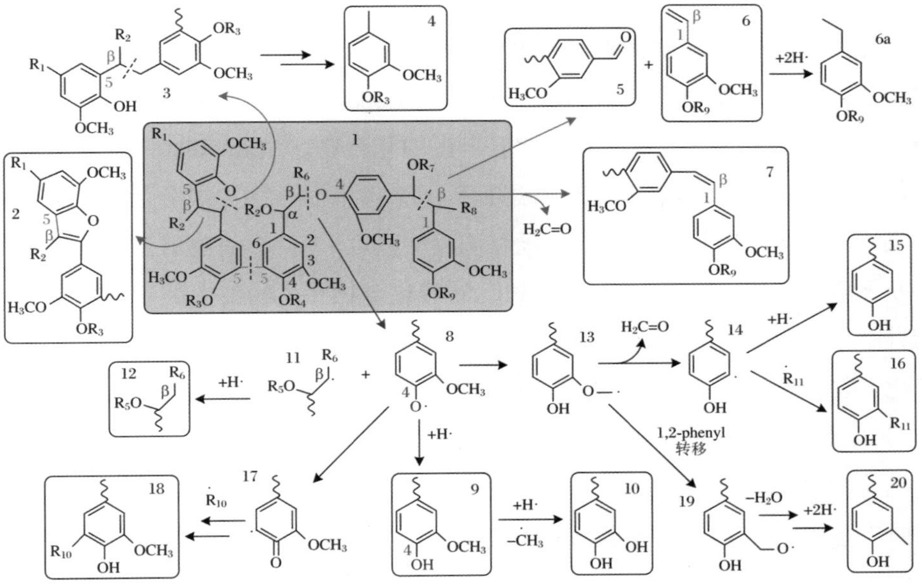

图1.4 木质素热解机理示意图

分子量各异的化合物。在热解过程中,生物油的产率受温度的影响最为明显。一般而言,在快速热解过程中,生物油的最大产率通常在热解温度为450~550 ℃时。例如,De Sisto 等在热解松木锯末时,获得的生物油的产率在500 ℃时达到最大值,并且生物油的成分中,羟基和甲氧基化合物的含量也明显高于在400 ℃和600 ℃热解得到的生物油的[48]。而在400 ℃热解时得到最高产率的热解生物炭,主要是因为低的热解温度容易导致生物质的不完全热解。而对于木薯秸秆和根茎的热解,达到油相的最大产率的温度分别为475 ℃和510 ℃,并且实验结果还显示,随着温度的上升,生物油的含水率是不断下降的,表明在低的热解温度下,有利于生物质组分发生脱水反应[86]。450~550 ℃作为生物质热解生成生物油的最佳温度区间的主要原因分析如下:在低的热解温度下(<450 ℃),生物质的挥发组分不容易完全分解,而在较高的温度下(>550 ℃),热解产生的挥发组分的二次分解将会主导热解过程,以降低生物油和热解生物炭产率的代价大量增加热解气体的产生量。研究人员在对纤维素热分解过程及其产物的分析中发现,温度在430~730 ℃之间时,热解气体的产率是一直增加的,而生物油的产率则是先增加,在570 ℃达到峰值,然后开始下降。TG-FTIR 的分析结果表明,热解气体形成的前体主要是羟基丙酮、羟基乙醛、5-羟甲基糠醛、糠醛和甘油醛。而这些化合物都是生物油的主要产物,这就表明,热解气体的来源大部分是生物质热解挥发产物的二次分解的产物,而不是生物质直接热解的产物[87],因此二次裂解反应对生物油产率的降低贡献远比对生物炭产率降低的贡献高[88]。

除了产率之外,热解温度对生物油组成的影响也是巨大的。[89]由于在低的温度下，生物质的分解主要发生在杂原子(N 和 O 的位点)上，因此，获得的生物油大部分是含有 N 和 O 官能团的化合物。例如，Fu 等在热解磷酸处理的木屑时发现，在低的热解温度下(350 ℃)，所得的生物油的主要成分包括 5-羟甲基糠醛、乙酸、羟基丙酮和羟基乙醛等含氧丰富的化合物。[90] Sanchez[91]等在研究市政污泥的热解时发现，在低的热解温度下，所得生物油的主要成分是羧酸、苯酚、吲哚等含 O 和 N 化合物，而其浓度在温度为 350~950 ℃ 时随着温度的升高而不断下降。相反，苯、甲苯、苯乙烯和吡啶等化合物在生物油中的含量随着温度的升高是不断上升的。这主要是因为随着热解温度的上升，一些不稳定的 N 和 O 化合物将会经历一个二次裂解的过程，产生更为稳定的气体物质 CO、CO_2、H_2O 以及碳氢化合物存在于生物油中，从而在高的温度下虽然获得生物油产率较低，但是通常该生物油具有较高的热值和稳定性。[92] Horne 等在热解废弃木屑时发现，550 ℃ 热解所得生物油中苯、甲苯、二甲苯、乙苯的含量远远高于其在 400 ℃ 热解所得生物油中的含量，而随着温度的进一步升高(>700 ℃)，生物油中多环芳烃芘、菲、蒽和萘等的含量又将会明显升高[93]。此外该研究还发现，随着温度的升高，生物油的含水率也将会明显上升。因此，温度的升高可以增强热解过程中的脱羰基、羧基和脱水反应，从而增加生物油中有机相的碳含量和稳定性，并降低其含氧量。

升温速率是另一个影响生物质热解过程的主要因素。高的升温速率可以导致生物质的快速分解，产生大量的挥发性物质。根据 Chan 等提出的机理[94]，生物质热解过程产物的分布情况主要取决于两个平行且竞争的反应发生：生物质分解产生挥发性化合物和生物炭，以及热解产生的挥发组分和生物炭的二次分解。在高的升温速率下，通常能产生更多的生物油组分，主要原因是热解组分的二次分解受到限制。因此，对于生物质的热解液化生产生物油的过程而言，高的升温速率是必要的，因为高的升温速率容易克服热解过程中传热限制，同时通过快速绝热分解生物质的过程增加挥发组分的丰度，从而减少热解过程中的二次反应的发生，诸如焦油的分解和重聚的发生[95]。升温速率的不同也可能改变获得最大生物油产率的最佳温度。研究人员在以不同的升温速率热解细茎针草生物质时获得的结果表明，在升温速率为 50 ℃·min^{-1} 和 150 ℃·min^{-1} 时，生物油的最大产率分别为 45% 和 57%，此时的温度为 500 ℃。而当升温速率达到 250 ℃·min^{-1} 时，生物油的最大产率达到 70%，而最佳温度也同时上升到了 550 ℃。[96]这种现象的主要原因可以解释为：在高的升温速率下，生物质热解过程通常能产生相对更多的挥发性物质，同时，也可以在更大的温度范围内发生分解。此外，在高的升温速率下，生物质床层的温度升高速率远大于挥发性物质的离开速率。

升温速率作为影响热解产物的产率分布的一个重要参数,在文献中也有大量的研究。[88,97-99]例如,Uzun 等在热解豆饼生物质过程中发现,慢速热解(升温速率:5 ℃·min^{-1})时生物油的产率比快速热解(升温速率:5 ℃·min^{-1})时获得的生物油的产率要低 23.4%,同时,生物油中的含水率随着升温速率的增加有明显的下降。[88] Angın 在研究升温速率对红花种子压饼生物质热解生产生物炭的过程中发现升温速率的增加可以降低生物炭的产率,同时也抑制了生物炭孔径的生长和比表面积的增加,但是对生物炭的元素组成影响并不是很明显。[100] Ozbay 等在研究不同升温速率对热解棉籽饼生物质的影响时发现,在升温速率为 0~300 ℃·min^{-1}时,生物油产率的增加值远高于其在升温速率为 300~700 ℃·min^{-1}时的增加值。[101] 类似的现象在热解木屑生物质的实验中也有发现,如 Salehi[98]发现,热解木屑生物质中生物油产率的增加值在升温速率为 500~700 ℃·min^{-1}时明显大于升温速率在 700~1000 ℃·min^{-1}时的增加值。

除了温度和升温速率这两个因素之外,还有其他一系列的因素可以影响生物质的热解过程,包括生物质进料的尺寸[102-104]、蒸气停留时间[105-107]、热解载气的种类和流速[108-110]、生物质的种类和组成(含水率、金属离子和无机成分的存在)[111-113]等。

1.4 生物质热解产物的品质优化提升

1.4.1 生物油

碳(C)、氧(O)和氢(H)是木质纤维素生物质的主要化学元素。典型的木质纤维素生物质通常含有 90% 以上的 C,H 和 O 元素(干基),这些元素是构建纤维素、半纤维素和木质素,以及其他必要的有机化合物,如核酸、蛋白质和激素等的主要成分。植物主要通过光合作用从生长环境中的 CO_2 和 H_2O 中同化 C,H,O 元素。生物质中 C,H,O 元素的相对含量受生长环境(如温度、水、土壤和空气)的影响显著。生物质热解过程中,C,H,O 元素主要转化为生物油、生物炭及焦油等组分,以下分别加以介绍。

生物质热解过程产生的生物油通常是棕黑色的黏稠液体,它由和生物质类似的元素组成。生物质的成分非常复杂,包含一系列含氧有机化合物(醇、醛、酚和羧酸等)以及水分(来自生物质本身或者热解过程发生的脱水反应产生的)等(图 1.5)[44],还有一些由于木质素、纤维素和半纤维素等生物质组分不完全热解产生的寡聚体,及一些由于过滤不完全而落入生物油中的灰分。生物油中的这些成分大部分都不是很稳定,反应活性很高,在其使用过程中存在诸多不足。表 1.2 归纳了生物油的一些缺点,以及其产生的原因和可能的影响。

表 1.2 生物油的一些缺点、产生原因和可能的影响

缺点	产生原因	可能的影响
强酸性(低 pH)	纤维素、半纤维素分解过程中产生羧酸	腐蚀性强
老化	醛、酮类物质的高反应活性,容易发生二次聚合	增加生物油的黏度,导致相分离
含有碱金属化合物	固液分离不完全	导致在生物油催化提质过程中催化剂中毒;在燃烧过程中容易造成固体的沉积;增加生物油的腐蚀性
存在固体物质(热解生物炭)	固液分离不完全	促进生物油的老化;导致生物油的相分离;导致催化剂中毒
含有氯、硫、氮元素	生物质进料被污染	导致在生物油催化提质过程中催化剂中毒;增加生物油的腐蚀性;产生异味
可蒸馏性能差	含有大量具有热不稳定性的化合物	降低生物油的可分离特性;增加从生物油中提取化学品的成本
黏度高	生物油中含有大量未完全热解的纤维素、半纤维素和木质素寡聚体	生物油的流动性差
含水率(含氧量)高	生物质进料中的水分和热解过程中的脱水反应,使生物油含有很多水	降低生物油的热值和稳定性,增加其腐蚀性;影响催化剂的性能
温度敏感性	含有大量具有热不稳定性的化合物	降低生物油的高温可操作性;容易产生相分离和老化
毒性	生物油的组成化合物中含有一些毒性较大的化合物	毒性虽然不大,但是不可忽略

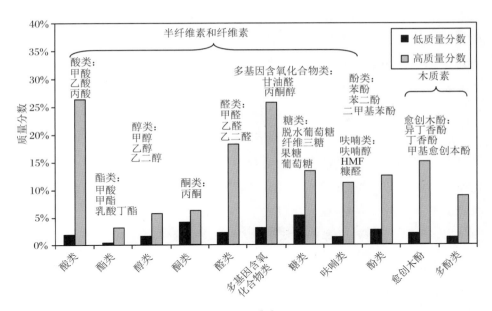

图 1.5　生物油中存在的典型有机物及其含量[44]

在表 1.2 中列出的生物油的缺点中,对生物油的燃料特性和与其他燃料的兼容性影响较大的因素包括高的含氧量(含水率)、固体(热解生物炭)的存在、高黏度以及化学不稳定性。因此,提升生物油的品质就是降低以上几个因素的影响。目前生物油的提质方法有多种,包括物理方法和催化化学方法。[39]

用于生物油提质的物理方法主要有过滤、乳化和溶剂添加法等[44],其中应用较为广泛的过滤方法主要包括高温蒸汽过滤和膜过滤。高温蒸汽过滤可以将生物油中的灰分含量降低至 0.01% 以下,而碱金属含量降低至 10 mg·kg^{-1},从而获得较高品质的生物油。然而,在热解过程中,灰分中含有的无机成分通常具有一定的催化活性,可以促进生物质的分解,因此,高温蒸汽过滤方法在去除灰分的同时,也将在一定程度上降低生物油的产率。Pattiya 和 Suttibak 在热解木薯生物质时,发现经过高温蒸汽过滤获得的生物油,尽管产率有所降低(大约降低 5%),但是其品质有了明显的提升(灰分含量从 0.20% 降低至不足 0.01%,动力黏度从 18 cSt 降低至 5.1 cSt,稳定性也有了显著的提升)。[114] Chen 等比较了经过高温蒸汽过滤和未经过此过程获得稻壳生物油的品质,发现经过高温蒸汽过滤的生物油中碱金属含量下降十分显著,其黏度也有一定的降低,但是生物油的含水率及 pH 呈现出升高的趋势。[115]

膜过滤方法可以将高温蒸汽过滤过程中无法过滤的微粒(粒径小于 10 μm)进行分离,而这些细小的微粒通常是造成生物油不稳定、易老化的主要因素之一。[116] Javaid 等使用孔径为 0.5~0.8 μm 的微孔陶瓷滤膜对快速热解的生物油进行微滤,结果表明,过滤之后生物油中的微粒数量大大降低,但是生物油作为能源应用的主要成分几乎没有改变[117]。膜过滤不仅可以将生物油中的固体微粒分离出来，还可以根据其组分中不同化合物的渗透压的差异，从生物油中分离出某一个组分。例如，Teella 等通过纳滤和反渗透膜从生物油的水相组分中分离得到了乙酸[118]。

生物油不能和烃类燃料很好地混合，但是在表面活性剂或者其他辅助条件的存在下，它可以和柴油通过乳化而混合。经过表面活性剂辅助的微乳化过程，可以将 5%～95% 的生物油和柴油混合形成乳液[119-120]。Guo 等通过超声和机械辅助的方法将生物油和柴油混合，获得了稳定性很好的乳液，最长可以稳定 31 天。[121] 机理分析表明，在乳化过程中，生物油中的小分子含氧化合物被油水膜包裹，形成一个特殊的球形壳状结构，通过这个独特的球形壳状结构，生物油在柴油中呈现出优异的乳化性能，而在超声和机械辅助下可以降低这个球形壳状结构的尺寸，从而增加乳液的稳定性。尽管乳化法可以很好地将生物油添加到柴油中作为燃料使用，然而乳化过程需要添加大量的表面活性剂，通常会大大增加成本，而使用超声或者机械辅助的方法耗能巨大，因此，乳化法的应用并不十分广泛。

在黏度很高的生物油体系中添加一定量的极性有机溶剂，可以增加生物油的流动性，降低其黏度。有机溶剂的添加，如甲醇、乙醇、甘油等，可以有效地增加生物油的稳定性。Diebold 等在生物油中添加 10% 的甲醇之后发现其老化速率降低了将近 20 倍。[122]Deng 等在生物油中添加一定量的甘油之后再进行蒸馏，可以将生物油中 95% 以上的组分回收，并且蒸馏过程中生物油的老化过程几乎可以忽略不计，而且甘油还可以多次回收利用。[123]

生物质本身的成分中含有一些具有一定催化活性的物质，它们包括碱金属和碱土金属的化合物，是作为生物质生长过程中的主要营养元素而存在于生物质的结构之中的。其中钾、钠元素含量最高，在生物质的热解过程中，它们可以作为催化剂促进热解蒸气的二次分解产生小分子气体，从而降低生物油的产率。因此，为了降低生物质本身的无机成分对生物质热解的副作用，在实际操作过程中，可以先通过一个简单的酸洗过程，将其中的无机成分清洗去除，从而降低其对热解蒸气的二次裂解的催化作用。

生物质品质提升至一定要求后可以作为交通燃料，包括柴油、汽油、煤油以及天然气等，通常会涉及加氢、脱氧和裂解等一系列的化学反应[38]，这些过程可以通过集成催化、热解和化学反应的方法来实现，主要过程有如下几种：催化加氢脱氧反应、酯化反应、催化热解蒸气裂解、通过费托合成方法生产小分子烃类物质等，具体的过程归纳如图 1.6 所示。

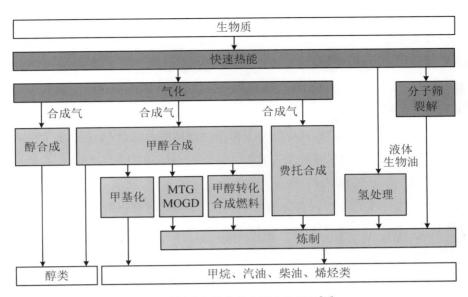

图 1.6　生物油品质提升至燃料及化学品的主要方法归纳[44]

催化氢化处理是生物油品质提升的主要的反应之一[124-128],它主要是将生物油中的氧和外加的氢气转化成水,从而达到降低生物油中氧含量的目的。氢化过程和快速热解过程通常是两个分离的过程,它的主要处理对象是液态的生物油,反应通常发生在高温(100 ℃ 以上)高压(>1 MPa)的条件下,并且需要额外的氢气。生物油完全氢化后将获得类似汽油的产物,经过一定的常规炼制过程之后就可以得到交通燃料了。催化氢化过程最主要的影响因素是催化剂,最早使用的催化氢化催化剂是将硫化的 CoMo 或者 NiMo 化合物负载于氧化铝或者硅铝酸盐载体上制备而成[129-132]的,它们的催化原理和石油组分的脱硫过程类似。但是,在应用过程中,这些催化剂将会产生一系列的问题,包括催化剂的载体氧化铝和硅铝酸盐在生物油的高含水率的条件下稳定性不高,催化剂中的硫元素在反应过程中容易流失,从而导致催化剂的迅速失活。随后,多孔材料(分子筛、活性炭等)负载的贵金属(Pd,Pt,Ru 等)催化剂开始广泛应用于生物油及其模型化合物的氢化过程中[133-136]。这些催化剂同时具有高稳定性和高活性的特点,可以很好地对生物油及其模型化合物进行氢化处理。同时由于大部分的贵金属对氢气具有很高的活化能力,因此,这些催化剂参与的氢化反应所需的氢气压力和反应温度要大大低于早期的硫化 CoMo 或者 NiMo 催化剂,从而降低反应的设备依赖性和操作难度。但是,氢化过程中的催化剂中毒是一个不可避免的大问题,生物油中存在的少量硫、氮、氯等元素在高温的条件下很容易导致贵金属催化剂因中毒而失活[38]。此外,贵金属催化剂的高成本也在一定程度上限制了它的广泛应用。

在催化氢化处理过程中,生物油通常会发生以下的变化:pH 增加、含水率增加、含氢量增加、动力黏度降低、含氧量降低,从而导致生物油的有机组分增加、稳定性增加、热值增加,更加接近交通燃料的品质[137-138]。最近,一个新型加氢酯化反应被开发出来,用于将生物油中的有机酸和醛酮类化合物转化成稳定的酯类化合物[139]。用于加氢酯化反应的催化剂通常具有双功能,例如酸性分子筛负载的贵金属纳米颗粒,其中的分子筛作为固体酸,可以催化有机酸和醇类的酯化反应,而贵金属纳米颗粒作为氢气的活化剂,可以催化醛和酮类化合物的加氢反应。Tang 等在 SBA-15 的分子筛修饰上磺酸基团,制成一个具有较强酸性的固体酸,然后再负载上 Pt 纳米颗粒,合成了双功能的催化剂,将其用于对生物油模型化合物乙醛和乙酸的加氢酯化反应。[134]结果表明,金属催化位点和酸性催化位点之间存在一定的协同作用,最后以较高产率获得最终的酯化产物乙酸乙酯。除了以上的例子,表 1.3 还归纳了不同催化剂用于生物油及其模型化合物的催化氢化反应。

表 1.3　不同催化剂用于生物油及其模型化合物的催化氢化反应

催化剂	生物油及其模型化合物的种类	反应条件	氢化作用效果	参考文献
含 Ru 的 Shvo 催化剂	白杨木热解生物油	氢气压力为 1 MPa,温度为 145 ℃,催化剂:反应物=200:1	醛类化合物含量显著下降	[140]
碳负载的 Pt,Pd,Ru	桉树木热解生物油	氢气压力为 2.1 MPa,温度为 320 ℃,反应时间为 4 h	氧含量下降明显,含水率上升	[141]
Ni-Co-Pd/γ-Al_2O_3	小球藻热解生物油	温度为 150~350 ℃,氢气压力为 2 MPa	含水率上升,热值增加,黏度下降	[142]
碳负载的 Ru	木质素热解生物油	温度为 300 ℃,氢气压力为 14 MPa	重质生物油转化为汽油类似产物	[143]
水溶性 Ru 催化剂	二氯甲烷萃取生物油	温度为 45~70 ℃,氢气压力为 4.5 MPa	羰基化合物含量显著下降	[144]
P/Mo 修饰的 NiCu/SiO_2-ZrO_2	愈创木酚(生物油模型化合物)	温度为 320 ℃,氢气压力为 11 MPa	生成各种加氢脱氧的产物,其中环己烷产率最高	[145]
Ni/HZSM-5	苯酚(生物油模型化合物)	温度为 160~240 ℃,氢气压力为 4 MPa	生成各种加氢脱氧的产物,其中苯的产率最高	[146]
碳负载的 Pd	黄杨木热解生物油	温度为 250~370 ℃,氢气压力为 3 MPa	含氧量下降,氢碳比上升	[147]

续表

催化剂	生物油及其模型化合物的种类	反应条件	氢化作用效果	参考文献
NiMo/Al$_2$O$_3$	松木热解生物油	温度为400 ℃,氢气压力为13 MPa	含氧化合物含量显著下降,热值达到46 MJ·kg^{-1}	[148]
NiCe/Al$_2$O$_3$ 等	麻风树热解生物油	温度为400～600 ℃,氢气压力未知	生物油品质明显提升	[149]
HY 分子筛和Al$_2$O$_3$ 负载 Pd	苯酚(生物油模型化合物)	湿度为250～350 ℃,氢气压力为1.5 MPa	苯酚几乎完全转化为环己烷和环己烯	[150]

催化氢化过程中,除了使用外加氢气源作为氢化反应物之外,也可以使用甲酸等物质,在反应过程中原位产生氢气对生物油及其模型化合物进行氢化。由于原位产生的氢通常比外加氢气具有更高的活性,因此,这一类反应通常能够在较为温和的条件下进行,产率和选择性都很高。Xiong 等采用分子筛负载 Pd 和 Ru 为催化剂,在甲醇溶剂中以甲酸作为氢源对稻壳热解生物油进行氢化处理,并将获得的结果与外加氢气源的氢化效果进行对比。[137]结果表明,甲酸原位氢化尽管反应较为温和,但是对生物油品质提升的效果和外加氢气的氢化效果相当。Deng 等通过酸水解生物质来源的碳水化合物(纤维素、淀粉和葡萄糖等)获得甲酸和乙酰丙酸,然后以 RuCl$_3$ 作为催化剂,通过甲酸分解原位产生的氢对乙酰丙酸进行氢化,获得 γ-戊内酯,可以作为航空燃料的替代品。[151] γ-戊内酯的最高产率能够达到 90%以上,反应选择性也很好。

和催化加氢处理不同,催化热解是在源头上提高生物油的品质[152-155],它和生物油的产生过程耦合在一起,通过在热解过程中加入特定性能的催化剂,催化热解过程中生物质的裂解过程,使得热解蒸气中某一类化合物的产生反应被增强,从而获得更高品质的生物油[156]。由于热解过程通常是在温度较高的情况下进行的,因此,用于催化热解的催化剂必须具有很高的热稳定性。此外热解产生的挥发性物质十分复杂,这就要求催化剂具有广泛的适应性。多孔分子筛催化剂具有很高的热稳定性、广泛的适应性,因此被广泛地作为生物质热解的催化剂。Carlson 等在三个不同的反应器中,使用 ZSM-5 型分子筛作为催化剂对松木锯末进行热解,发现催化热解所产的生物油中芳香烃类和烯烃类的含量显著增加。[157]在催化热解过程中,为了保证热解蒸气进入催化剂孔隙中发生催化反应,避免其直接发生二次分解,一般要求升温速率很高且催化剂用量很大。[158] Zhang 等在催化热解松木和醇类(甲醇、1-丙醇、1-丁醇、2-丁醇)混合物时,使用高达 40%的 ZSM-5 作为催化剂,获得了含有大量 C$_2$～C$_4$ 烯烃类及芳香烃类化合物的生物油。[159]结果表明,600 ℃是最佳的催化热解温度。除了 ZSM-5 分子筛之外,其他分子筛催化剂也广泛应用于生物质的催化热解过程中。表 1.4 归纳了不同催化剂对生物质快速热解过程的催化作用及效果。

表 1.4 不同催化剂对生物质快速热解过程的催化作用及其效果

催化剂	生物质种类	反应条件	催化效果	文献
CaO/MgO	松木	反应温度为 350 ℃,催化剂与进料比最少为 1:3,最多为 3:1	生物油酸性和含氧量显著下降,热值上升	[160]
ZSM-5	稻壳	反应温度为 350~600 ℃,催化剂与进料比最少为 1:2,最多为 2:1	生物油的含氧量显著下降,酚类化合物有含量有所升高	[161]
Mo_2N/Al_2O_3	木质素	反应温度为 700~850 ℃,催化剂与进料比为 4:1	苯的产率最高达 70.1%	[162]
CaO/Al_2O_3/高岭土	Karanja seed	反应温度为 550 ℃,催化剂与进料比为 2:1 至 8:1	获得了富含烃类化合物的生物油	[163]
ZSM-5	松木、玉米芯、秸秆等	反应温度为 600 ℃,催化剂与进料比为 9:1	芳香化合物的产率达到 38.4%	[164]
Y 型分子筛	白杨木	反应温度为 400~600 ℃	生物油的有机组分增加,pH 增加,黏度降低	[165]
Y 型和 β 型分子筛	橡木	反应温度为 400~600 ℃	生物油的 C/O 比从 1.9 上升至 5.8	[166]
ZSM-5	生物质和塑料混合物	反应温度为 400~650 ℃	芳香化合物的产率和选择性提高很大	[167]
SBA 负载的 Pd	杨木	反应温度为 600 ℃	热解木质素转化为酚类物质	[168]
HZSM-5 分子筛	松木	反应温度为 650 ℃,催化剂与进料比为 9:1	42.5%的生物质碳元素转化为烃类化合物,其中芳香烃类选择性最高	[169]
HZSM-5 分子筛	松木	反应温度为 400~650 ℃,催化剂与进料比为 4,9,11:1	芳香烃类化合物的含量显著增加	[170]

催化热解过程需要在一个特定的高温下进行,这就要求所用的催化剂具有高的热稳定性和机械稳定性。构建一个合适的催化体系需要的条件十分苛刻,而且催化热解过程中不可避免地会发生结焦问题,从而会大大降低催化剂的反应活性,使催化剂的再生难度增加和循环使用性能降低。由于存在以上不足之

处,催化热解过程并没有被广泛推广为一个成熟的生物油品质提升的方法。

除了以上提到的应用最为广泛的催化氢处理和催化热解方法,目前报道的生物油品质提升的方法还包括:① 水相蒸气重整[171-174]:从生物油的水相成分,主要是烃类含氧衍生物(醛、酮、酚类)经过一系列的脱水、脱羧和氢化反应,生成氢气、一氧化碳和烷烃类化合物;② 费托合成[175-178]:先将生物质或者生物油气化转变为合成气,然后在催化剂的作用下,将合成气转化为小分子的烃类化合物,可以作为汽油的替代。

1.4.2 生物炭

生物质因其环境友好性、可再生性以及产量巨大等特点,成为一种合成功能碳材料的理想原材料[179]。以生物质为原料制备的功能碳材料在污染物的去除、催化、储氢、储能等领域都有广泛的应用。从生物质到功能碳材料,碳化是一个必不可少的步骤,它主要是将生物质中的挥发组分除去,留下富碳的骨架结构。另一个步骤是活化,将从生物质获得的碳材料根据应用目的的不同,通过一系列特定的物理或者化学过程,对其进行结构的修饰,使其满足应用的要求。其中碳化步骤是最主要的步骤,目前应用较为广泛的生物质碳化过程包括水热碳化和惰性气体氛围下热化学碳化。

水热碳化是指以水作为溶剂,将生物质置于高压反应釜中在高于 100 ℃ 的温度下加热,利用水在高温下自身产生的高压使得生物质发生碳化,来合成碳材料[179]。水热碳化过程通常发生在温度低于 300 ℃ 的条件下,通常会有脱水、聚合、缩合等一系列反应发生,生物质水热碳化的过程和自然界煤的形成过程十分类似,不同的是煤的形成需要极其漫长的地质变化才行,而生物质的水热碳化可以在很短的时间内完成[180]。水热碳化过程是一个自发的绝热过程,因此生物质中的碳大部分将最终转化为碳材料中的碳,额外的损失很少,具有很高的碳原子效率,符合原子经济性原理。在过去的几十年间,大量的生物质基水热碳材料被合成出来,并且在很多领域都呈现出很好的应用前景[181]。而随着近来人们对水热过程的深入研究以及机理的阐述[182],水热碳化反应已经开始用于在生物质基础上设计新型的具有特殊功能的碳基材料了。水热碳材料可作为污染物吸附、催化剂载体以及电容器类储能材料等[183]。

由于生物质的水热碳化过程具有操作简便、成本较低并且容易控制的优点,因此目前这方面的研究成果层出不穷。中国科学技术大学的俞汉青课题组分别在 2008 年和 2010 年综述了通过生物质水热碳化过程合成碳基功能材料的进展[179,181]，包括水热碳化方法对碳材料的表面调控过程和机理，碳和其他材料在水热过程中的自组装和复合过程等，都进行了较为深入的分析。作者还详细综述了生物质来源的水热碳及其复合材料在不同领域的应用情况，包括在环境、催化、电子和生物医学等方面。2012 年，德国 Max-Planck 胶体与界面研究所的 Titirici 课题组展望了水热碳化过程的未来发展趋势[184]，作者主要分析了该过程中碳的形成机理，展望了水热碳的大规模生产、纳米构建、功能化以及应用方面的趋势和存在的问题。他们指出，未来水热碳化过程要实现低成本、可持续的操作，使用生物质作为碳源是一个理想的举措，而实现水热碳材料的高效、低成本的分离也是未来进行这方面研究必须要解决的一个主要问题。

和水热碳化方法不同，热化学碳化过程通常发生在更高的温度（大于 500 ℃）下，碳化过程中生物质的挥发组分是在惰性气体的夹带下挥发出去的，由于其中含有大量的含碳组分，因此热化学碳化过程的碳原子效率显著降低。但是水热碳材料通常含有过多的含氧功能基团，相对而言，热化学碳化的材料的其他杂原子含量将会大为降低，其含碳量明显上升，因此其导电性等特性也明显增强。在热化学碳化过程中，碳化温度是影响材料结构和性质最主要的因素[185]，一般而言，高的碳化温度（600 ℃以上）可以增加挥发性物质的释放，降低碳材料的产率，但是挥发性物质的释放会使剩下的碳材料孔隙变得丰富，比表面积增大。升温速率是另一个重要的影响因素，低的升温速率有利于生物质发生脱水反应，以及其结构中的聚合物链的稳定[186-187]，因此，可以使得碳材料的产率更高。但是升温速率对碳材料的孔隙结构的影响不大。

直接碳化制备的碳材料通常比表面积不是很大，孔隙结构也不是很丰富，并且表面功能基团也不是很多，因此需要进一步通过活化过程来改善其特性，从而使其具有很好的使用性能。活化的主要目的是增加碳材料的孔容积，增加其多孔性结构和比表面积。活化过程主要包括物理活化和化学活化，物理活化通常使用水蒸气和 CO_2 作为活化试剂[188-189]，而化学活化则是采用化学试剂对材料进行处理，常用的活化试剂有 KOH、$ZnCl_2$、H_3PO_4、$NaOH$ 等[190-193]。

2013 年，Nor 等综述了以生物质为原料，经过热化学碳化和活化方法制备不同的活性炭材料[185]。作者归纳了热化学过程的主要影响因素以及通过不同活化方法制备的碳材料的组成和结构特征，并且详细综述了生物质来源的活性炭在空气污染物去除方面的应用，包括去除 SO_2、NO_x、H_2S 以及挥发性有机物（VOCs）等。而在稍早一些的 2009 年，Demirbas 综述了以农业废弃生物质为原料通过热化学碳化制备活性炭材料的方法,并将其用于吸附去除染料废水中染料[194]。

由于其多孔性和化学稳定性,活性炭不仅可以用作吸附材料,还可以作为载体负载其他的活性物质,例如金属纳米颗粒或者氧化物等,或者在其表面修饰上一些具有催化作用的其他基团,例如 SO_3H,可用作催化剂材料。Arancon 等[195]以玉米芯生物质为原料,先在 400~600 ℃下碳化获得活性炭材料,然后再进行磺化,制备出多孔碳基固体酸材料。该材料具有较强的酸性,用于催化酯交换反应合成生物柴油时,其能表现出优异的催化活性,可以在 6 h 内将废弃油脂转化为脂肪酸甲酯,产率可达 95% 以上。2011 年,Balakrishnan[196]综述了以废弃材料作为原料合成催化剂材料,指出自然界广泛存在的植物生物质,如稻壳,可以作为催化剂或者载体材料,并且由于其在自然界中广泛存在,催化剂的合成成本十分低廉。

活性炭材料的多孔性结构、高比表面积、大的孔容积以及优良的导电性能,使其可以广泛地被用作超级电容器的电极材料,用于电化学储能。Biswal 等[197]以树叶为原料,通过直接的热解方法,不经过任何的物理或者化学活化方法,合成出一种具有高比表面积的微孔活性炭材料,并测试了其电化学储能性能。作者指出,合成的材料具有高的导电性和微孔率,而这两个因素是该材料具有很高的电容(400 $F \cdot g^{-1}$)和能量密度(55 $Wh \cdot kg^{-1}$)的主要原因。作者同时比较了不同生物质来源和活化方法制备的活性炭的电容性能,结果如表 1.5 所示。

表 1.5　不同生物质来源和活化条件制备的活性炭材料的电容性能比较

生物质	活化试剂	比表面积 ($m^2 \cdot g^{-1}$)	最大电容 ($F \cdot g^{-1}$)	电解液	参考文献
稻壳	NaOH	1886	210	3 $mol \cdot L^{-1}$ KCl	[198]
冷杉木	H_2O	1131	140	0.5 $mol \cdot L^{-1}$ H_2SO_4	[199]
开心果壳	KOH	1096	120	0.5 $mol \cdot L^{-1}$ H_2SO_4	[200]
冷杉木	KOH	1064	180	0.5 $mol \cdot L^{-1}$ H_2SO_4	[200]
竹子	KOH	1251	260	30% H_2SO_4	[201]
香蕉纤维	$ZnCl_2$	1097	74	1 $mol \cdot L^{-1}$ Na_2SO_4	[202]
玉米粒	KOH	3199	257	6 $mol \cdot L^{-1}$ KOH	[203]
废咖啡豆	$ZnCl_2$	1019	368	1 $mol \cdot L^{-1}$ H_2SO_4	[204]
废报纸	KOH	416	180	6 $mol \cdot L^{-1}$ KOH	[205]
海藻	无活化剂	746	264	1 $mol \cdot L^{-1}$ H_2SO_4	[206]
甘蔗渣	KOH	1788	300	1 $mol \cdot L^{-1}$ H_2SO_4	[207]
木薯皮	KOH	1352	311	0.5 $mol \cdot L^{-1}$ H_2SO_4	[208]
瓜子壳	KOH	2509	355	30wt% KOH	[209]

1.5 生物质热解过程中污染物的产生、迁移、转化过程与机制

1.5.1 热解焦油

焦油是生物质热解过程中不可避免的副产品，被认为是限制生物质热解技术商业化的主要瓶颈，因为它不仅可以污染和侵蚀热解设备，还造成生物质热解的有效能量的浪费，与绿色和可持续的生物质热解特性相矛盾。焦油是含有一百多种化合物的复杂混合物，主要包括苯、酚类、N-杂环化合物和多环芳烃等，与一些小颗粒杂质冷凝形成复杂的结构，堵塞在下游设备的管道、过滤器或换热器中，造成整个系统的机械故障，或使生物精炼过程中的催化剂失活。此外，焦油中的各种芳香类和 N-杂环化合物具有毒性，对环境危害极大。

如表 1.6 所示，焦油根据其外观和性质可分为一级、二级和三级焦油。根据热解反应器的设计、原料和操作条件的不同，生物质热解和气化的焦油产率在 $0.5 \sim 100 \, \text{g} \cdot \text{m}^{-3}$ 之间。焦油通常是在热解过程中形成的。一级焦油的主要成分为含氧有机物，随着热解温度的升高，含氧有机物可以转化为低碳烷烃、芳香烃和烯烃，随着温度的进一步升高，这些烃类又可以转化为较高的烃类和较大的多环芳烃（图 1.7）。

表 1.6 焦油的形态特征

类别	形态特征
一级焦油	低分子量的含氧有机物,包括葡聚糖、糠醛、羟甲基乙醛等,主要形成于 400~700 ℃
二级焦油	酚类和烯烃类化合物,包括苯酚、甲基苯酚、二甲苯等,主要形成于 700~850 ℃
三级焦油	复杂的芳香类化合物,包括苯、萘、嵌二萘、甲苯等,主要形成于 850~1000 ℃

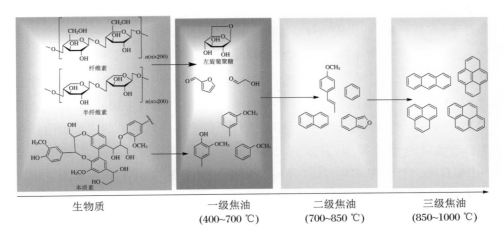

图 1.7 不同热解温度下焦油的形成过程分析

由于焦油是生物质热解过程中有害的副产物,因此如何有效地减少和去除焦油是一个值得关注的关键问题。在生物质热解过程中,有许多可以脱除焦油的方法。一般来说,这些方法可分为物理法(使用陶瓷过滤器或湿式洗涤器)和热化学转化法(通过高温裂解或使用催化剂将焦油转化为合成气)。物理方法只从气体产物中去除焦油,并未对其进行处置。和物理方法相比,热化学转化受到更多的关注,因为其可以将焦油转化为有用的气体产品,从而提高热解过程的整体效率。然而,仅仅通过高温热化学裂解焦油将大型有机分子分解成小的不凝性气体是没有吸引力的,因为它需要大量的能源输入,同时产生大量的烟尘。而催化热化学转化可以更有效地降低反应温度并高效地将焦油转化为有用气体,使其成为目前焦油去除研究的主要方向。焦油的催化转化主要有两种方法:一种是将催化剂与生物质原料混合,这样焦油就可以在热解反应器内转化;另一种方法是使用一个位于裂解反应器下游的分离反应器,将焦油转化。

1.5.2
持久性有机污染物

持久性有机污染物,如多环芳烃(Polycyclic Aromatic Hydrocarbons,PAHs)和含氧多环芳烃,以其高毒性以及对人类和野生动物的健康风险而闻名。燃烧生物质被认为是持久性有机污染物(Persistent Organic Pountants,POPs)产生的主要来源。相比之下,热解提供了一种相对环境无害的方法,因为在无氧条件下,有毒 POPs 的生成和释放将会显著减少。

多环芳烃是一类半挥发性化合物,具有极大的环境风险,与一系列健康问题有关,包括癌症、突变或人体畸形等。据估计,全球每年多环芳烃的排放总量超过 520 Gg,其中 80% 以上来自发展中国家,一半以上来自生物质燃烧[210]。例如,室内生物质燃烧(烹饪燃料)是多环芳烃所致室内空气污染的主要原因。在生物质燃料燃烧过程中,共发现 22 个原生多环芳烃、12 个硝基多环芳烃(nPAHs)和 4 个羟基多环芳烃(oPAHs),其中 nPAHs 和 oPAHs 主要来自环境空气中的二次生成,而多环芳烃主要来自生物质燃烧的一次排放[211]。

虽然在热解条件下可以显著抑制多环芳烃的生成,但生物质热解产物中仍存在一定数量的多环芳烃,尤其是焦油。Fagernas 等[212]分析了桦木生物质热解产物(气体、焦油、生物油)中多环芳烃的含量和分布,结果表明,多环芳烃主要集中在不凝性气体、挥发性可凝性液体、生物炭、焦油中,特别是在蒸馏釜底部收集的重焦油中。Zhou 等[213]研究了 9 种不同有机固体废物组分(木聚糖、纤维素、木质素、果胶、淀粉、聚乙烯(PE)、聚苯乙烯(PS)、聚氯乙烯(PVC)和聚对苯二甲酸乙二醇酯(PET))热解过程中 2-4 环多环芳烃的形成。结果表明,聚苯乙烯热解产生的多环芳烃最多,其次是聚氯乙烯(PVC)、聚酯(PET)和木质素,塑料的多环芳烃含量高于生物质。从多环芳烃的化学结构来看,焦油样品中多环芳烃的含量以 2 环多环芳烃为主,其质量分数为 40%~70%,热解过程中形成的 2 环多环芳烃中以萘最为丰富。3 环多环芳烃中以菲和芴最为丰富,而苯并[a]蒽和䓛在 PS、PVC 和 PET 的焦油中含量显著。

与焦油相和气相相比,固体生物炭相中也发现了大量的多环芳烃。以慢速热解获得的生物炭为例,多环芳烃总含量为 $0.07 \sim 3.27\ \mu g \cdot g^{-1}$,其含量随生物质来源的不同而变化。而快速热解获得的生物炭中多环芳烃总含量显著增加到 $45\ \mu g \cdot g^{-1}$。这一现象的主要原因可能是缓慢热解过程中产生的多环芳烃逸散到气相,而快速热解过程中产生的多环芳烃可能会在生物炭材料本身上凝聚。

表1.7总结了不同生物质和热解条件下生物炭中多环芳烃的浓度。

表1.7 不同生物质来源及生产条件下获得的生物炭中多环芳烃的含量比较[214]

生物质	生物炭制备方法	多环芳烃萃取方法	多环芳烃浓度($\mu g \cdot g^{-1}$)	参考文献
木屑（赤桉）、豌豆秸秆（豌豆）、植被残渣（主要是桉树的叶和枝）	木屑和秸秆等生物质在马弗炉中450℃热解1h	70 mL 庚烷超声萃取1h。萃取液通过0.5 μm PTFE 微孔纤维膜	0.07~0.12	[215]
日本柏木、日本栗子木、毛竹	生物质在马弗炉中400~800℃热解1h	通过3:7的甲苯/乙醇索氏提取16h	0.08~0.11 0.06~0.12 0.08~0.15	[216]
北美松树生物质	以不同的升温速度在450~1000℃下热解1h	根据国家标准分析方法进行分析	1.01~3.68	[217]
桦木及松木	以1.5~2℃·min^{-1}在600℃下热解	按照 Amtliche Sammlung §35 LMBG Method L07.00-40 分析、提取甲苯	0.005 0.013	[218]
桉树木质生物质、桉树叶	在400℃或550℃下慢速热解40 min	二氯甲烷索氏提取12h	忽略不计	[219]
树皮生物质	—	70 mL 甲苯索氏提取16h	45	[220]

有机固体废弃物热解过程中多环芳烃的形成机理有很多，其中去氢乙炔加成(Hydrogen-Abstraction/Acetylene-Addition, HACA)被广泛接受。根据这一机理，苯可以通过苯乙炔生成萘(式(1.1))，而苯乙炔在上述有机固体废弃物热解过程中产生的焦油中含量丰富。因此，单环化合物(如苯和苯乙烯)可以被认为是2环多环芳烃的主要前体。对于含有芳香环的生物质组分，如木质素等，芳香结构可能通过 HACA 机制生成2环多环芳烃。对于不含芳香环的组分，如纤维素、半纤维素等，多环芳烃主要通过 Diels-Alder 型反应生成烯烃和二烯烃。

$$\text{苯} \xrightarrow[-H_2]{HC\equiv CH} \text{苯乙炔} \xrightarrow{HC\equiv CH} \text{萘} \quad (1.1)$$

1.5.3 重金属污染

在环境污染日益严重的今天,植物的生长过程难免会受到各种污染物的侵蚀,导致收获的生物质中或多或少会含有一些污染物。例如,在重金属污染的土壤中生长的植物,将会从土壤中吸收大量的重金属,使植物生物质中的重金属含量显著上升,达到降低土壤中重金属污染的目的,这就是植物修复的主要原理[221]。由于植物修复方法操作简便,成本较低,因此广泛用于重金属污染的土壤的处理,效果明显[222-225]。但是,植物修复过程副产大量重金属污染的生物质,如果处置不当,其中的重金属将会再次释放出来,造成二次污染。快速热解作为一个生物质资源化的主要方法,对于处理重金属污染的生物质而言,是一个有意义的尝试。

Stals 等[226-227]将受重金属 Zn,Cd,Pb,Cu 污染的短期轮作作物和硬木进行快速热解,研究其中的重金属的迁移、转化和在热解产物中的分布情况。结果表明,在 350 ℃热解获得的生物油中重金属的含量很低(Cu 和 Zn 小于 5 mg·kg^{-1};Cd 和 Pb 小于 1 mg·kg^{-1}),大部分的重金属都集中在热解生物炭或者灰分中。而温度是影响重金属在热解产物中分布的主要因素:在 450 ℃以下热解,重金属主要集中在热解生物炭中,而当温度升高至 550 ℃时,挥发至生物油相中的重金属含量将会显著上升。作者还将热解生物炭中的重金属用不同的溶液进行浸出实验,发现大约 35%的重金属可以被 EDTA 等溶液浸出,这部分重金属被认为是植物可利用的,因此热解生物炭仍然不能直接处理至自然环境中,但是相比于原生物质而言,其体积和质量都大大下降,无害化处理相对简单,并且重金属的含量将会显著富集,可以通过较为简单的方法加以回收。与 Stals 同一个课题组的 Lievens 则研究了热解过程的影响因素,包括生物质种类、温度、传热载体等对热解过程中重金属的变化行为的影响[228-230]。结果表明,在较低的热解温度(<400 ℃)下获得的生物油组分几乎不含重金属成分,而在高温(>400 ℃)的情况下,重金属镉的挥发性明显高于铅、锌和铜。此外,锌和镉的挥发性还与生物质的种类密切相关。传热载体作为一个使生物质热解过程中受热更加均匀的辅助试剂,对热解过程中金属的变化行为也有一定的影响,泡沫二氧化硅载热剂对重金属铜和铅的保留能力要比沙子载热剂强很多。

植物修复过程获得的重金属污染生物质中重金属含量不可控,对研究热解过程中的重金属的变化行为会产生更多的不可控的变量。例如,重金属对热解的催化作用，不同种类和含量的重金属化合物的挥发性的变化等。因此，为了更好地研究重金属在生物质热解过程中的变化行为，Koppolu等[231-232]将未受污染的普通玉米秸秆生物质浸渍确定含量的重金属化合物(Ni,Zn,Cu,Co,Cr)制备成重金属污染的生物质，然后在不同的热解条件下对其进行热解，研究重金属对生物质热解过程的影响以及热解过程中重金属的迁移、转化和分布情况。结果表明，在实验室规模的热解条件下，98.5%的重金属都集中在热解生物炭固相之中，其重金属的浓度是热解进料中浓度的4～6倍。而在中试规模下，重金属在热解生物炭固相中的富集率仍然接近99%，其重金属的浓度是热解进料中浓度的3.2～6倍。同时，结果还显示，重金属的存在对提高热解过程生物油的产率和性质也有一定的帮助。

1.5.4
含氮污染物

氮元素是植物生长过程中的必需元素，植物通过吸收自然界中的氮元素来构建起本身的结构和功能，因此植物生物质中通常含有大量的氮元素。这些氮元素在热解过程中，经过一定的迁移转化过程，最终将变成对环境有害的氮氧化物污染物。最近，生物质热解过程中氮元素转化为氮氧化物前体的过程和原理通过多种研究方法得以阐述：Ren等[233]首先将氨基酸作为生物质中含氮化合物的模型，通过热重耦合傅里叶变幻红外光谱(TG-FTIR)来研究热解过程中氨基酸的变化行为。作者通过监测不同热解气氛下HCN,NH_3,HCNO,NO等氮氧化物的前体的生成情况，来阐述氮元素在热解过程中的变化行为。对于天门冬氨酸而言，在氧化氛围(95%氩气/5%氧气)下，HCN,HCNO,NO的释放量明显增加，而在惰性气氛(100%氩气)下，主要释放NH_3。而对于谷氨酸，HCN和HCNO的释放在氧化氛围(95%氩气/5%氧气)下是占主导地位的，但是NH_3和NO则是在弱氧化氛围(95%氩气/5%二氧化碳)下释放较多。对于芳香族的苯丙氨酸而言，其热解产物的释放规律较前两个也有很大不同，HCN和NH_3都是在惰性气体氛围(100%氩气)和弱氧化氛围(95%氩气/5%二氧化碳)下释放较多，而HCNO和NO则在惰性气体氛围(100%氩气)和氧化气体氛围(95%氩气/5%氧气)下释放较多。其中的原理主要是和不同氨基酸中的结构特征和其中氨基的稳定性有关。在以上研究的基础上，Ren等[234]继续将氨基酸和生物质的主要组成单元纤维素、半纤维素和木质进行共热解，研究氮元素在其中的变化行为。作者发现杂环氨基酸和脂肪族氨基酸的热解行为有明显的差异,而木质素可以促进氨基酸热解生成 NH_3,纤维素和半纤维素则抑制 NH_3 的产生。

和 Ren 等的研究工作相比,Becidan 等的工作则更进一步,他们直接研究含氮生物质热解过程中 NH_3 和 HCN 等氮氧化物的前体的产生情况[235]。作者研究了不同热解温度和升温速率下 NH_3 和 HCN 释放情况发现,在高的升温速率条件下 NH_3 是生物质热解的主要含氮产物,其产率随着热解温度的升高而不断增加,在 825～900 ℃ 之间达到峰值,生物质中氮元素转化为 NH_3 的转化率在 31%～38% 之间,而 HCN 的转化率则在 9%～18% 之间,也随着温度的升高而不断增加。在低的升温速率条件下,尽管 NH_3 仍然是主要的含氮产物,但是其转化率明显下降,只有 7%～19.2%。根据实验结果,作者提出了氮元素在热解过程中的转化机理,如图 1.8 所示。

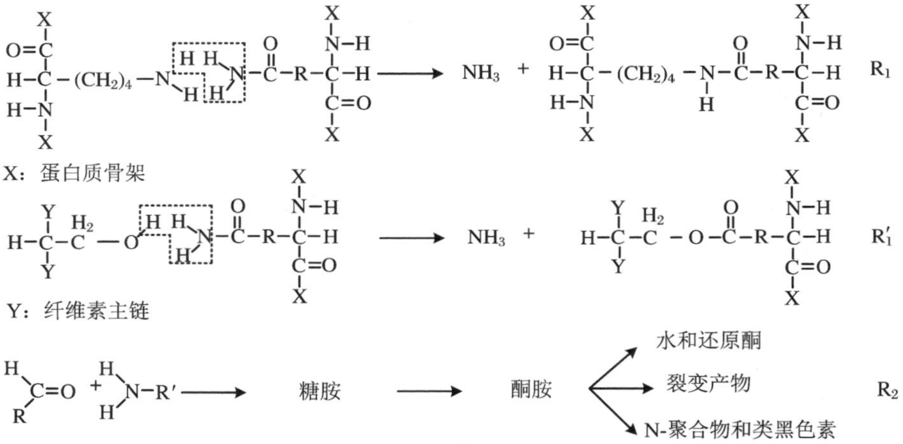

图 1.8 生物质在快速热解过程中氮元素的转化过程机理示意图[235]

Monash 大学的 Li 课题组研究了生物质和煤在共热解过程中氮元素和硫元素的迁移转化过程和一系列影响因素[236-239]。他们提出:NH_3 主要是由热解过程产生的 H 自由基对生物质中的含氮位点进行氢化形成的,而那些不稳定的杂环结构则是形成 HCN 的主要来源。而升温速率则通过影响 H 自由基的产生来影响 NH_3 的生成情况,在高的升温速率下,H 自由基的产生更为迅速,从而使得 NH_3 在高的升温速率下更容易形成。以上机理也可以解释,NH_3 容易在惰性热解气氛中生成,而在氧化性气氛中,H 自由基的产生受到抑制,导致 NH_3 的产量大为降低。而在水蒸气存在的情况下,由于产生 H 自由基的来源大为增加,因此产生的 NH_3 的量也明显增加。

参考文献

[1] Department of Environmental Protection, People's Republic of China. The state of the environment in China (in Chinese) [R]. 2011.

[2] Liao Y, Koelewijn S-F, Van Den Bossche G, et al. A sustainable wood biorefinery for low-carbon footprint chemicals production [J]. Science, 2020, 367(6484): 1385-1390.

[3] Shen G, Tao S, Wang W, et al. Emission of oxygenated polycyclic aromatic hydrocarbons from indoor solid fuel combustion [J]. Environmental Science & Technology, 2011, 45(8): 3459-3465.

[4] Atkins A, Bignal K L, Zhou J L, et al. Profiles of polycyclic aromatic hydrocarbons and polychlorinated biphenyls from the combustion of biomass pellets [J]. Chemosphere, 2010, 78(11): 1385-1392.

[5] Mandalakis M, Gustafsson Ö, Alsberg T, et al. Contribution of biomass burning to atmospheric polycyclic aromatic hydrocarbons at three European background sites [J]. Environmental Science & Technology, 2005, 39(9): 2976-2982.

[6] Tian D, Hu Y, Wang Y, et al. Assessment of biomass burning emissions and their impacts on urban and regional PM 2.5: a georgia case study [J]. Environmental Science & Technology, 2008, 43(2): 299-305.

[7] Obrist D, Moosmüller H, Schürmann R, et al. Particulate-phase and gaseous elemental mercury emissions during biomass combustion: controlling factors and correlation with particulate matter emissions [J]. Environmental Science & Technology, 2007, 42(3): 721-727.

[8] Szidat S. Sources of Asian Haze [J]. Science, 2009, 323(5913): 470-471.

[9] http://hjj.mep.gov.cn/stjc/201107/P020110712332691223881.pdf.

[10] Shen F, Xiong X, Fu J, et al. Recent advances in mechanochemical production of chemicals and carbon materials from sustainable biomass resources [J]. Renewable and Sustainable Energy Reviews, 2020, 130: 109944.

[11] Yu H, Huang G H. Effects of sodium acetate as a pH control amendment on the composting of food waste [J]. Bioresource Technology, 2009, 100(6): 2005-2011.

[12] Lesteur M, Bellon-Maurel V, Gonzalez C, et al. Alternative methods for determining anaerobic biodegradability: A review [J]. Process Biochemistry, 2010, 45(4): 431-440.

[13] Nallathambi Gunaseelan V. Anaerobic digestion of biomass for methane production: A review [J]. Biomass and Bioenergy, 1997, 13(1/2): 83-114.

[14] Themelis N J, Ulloa P A. Methane generation in landfills [J]. Renewable Energy, 2007, 32(7): 1243-1257.

[15] Dodds D R, Gross R A. Chemistry: Chemicals from Biomass [J]. Science, 2007, 318(5854): 1250-1251.

[16] Vispute T P, Zhang H, Sanna A, et al. Renewable chemical commodity feedstocks from integrated catalytic processing of pyrolysis oils [J]. Science, 2010, 330(6008): 1222-1227.

[17] Ate F, Miskolczi N, Borsodi N. Comparision of real waste (MSW and MPW) pyrolysis in batch reactor over different catalysts. Part I: Product yields, gas and pyrolysis oil properties [J]. Bioresource Technology, 2013, 133: 443-454.

[18] García A N, Font R, Marcilla A. Kinetic study of the flash pyrolysis of municipal solid waste in a fluidized bed reactor at high temperature [J]. Journal of Analytical and Applied Pyrolysis, 1995, 31: 101-121.

[19] Li A M, Li X D, Li S Q, et al. Pyrolysis of solid waste in a rotary kiln: Influence of final pyrolysis temperature on the pyrolysis products [J]. Journal of Analytical and Applied Pyrolysis, 1999, 50(2): 149-162.

[20] Velghe I, Carleer R, Yperman J, et al. Study of the pyrolysis of municipal solid waste for the production of valuable products [J]. Journal of Analytical and Applied Pyrolysis, 2011, 92(2): 366-375.

[21] Buah W K, Cunliffe A M, Williams P T. Characterization of products from the pyrolysis of municipal solid waste [J]. Process Safety and Environmental Protection, 2007, 85(5): 450-457.

[22] Baggio P, Baratieri M, Gasparella A, et al. Energy and environmental analysis of an innovative system based on municipal solid waste (MSW) pyrolysis and combined cycle [J]. Applied Thermal Engineering, 2008, 28(2/3): 136-144.

[23] Luo S, Xiao B, Hu Z, et al. Effect of particle size on pyrolysis of single-component municipal solid waste in fixed bed reactor [J]. International

Journal of Hydrogen Energy, 2010, 35(1): 93-97.

[24] Kumar R, Strezov V, Weldekidan H, et al. Lignocellulose biomass pyrolysis for bio-oil production: A review of biomass pre-treatment methods for production of drop-in fuels [J]. Renewable and Sustainable Energy Reviews, 2020, 123: 109763.

[25] Kageyama H, Osada S, Nakata H, et al. Effect of coexisting inorganic chlorides on lead volatilization from CaO-SiO_2-Al_2O_3 molten slag under municipal solid waste gasification and melting conditions [J]. Fuel, 2013, 103: 94-100.

[26] Björklund A, Melaina M, Keoleian G. Hydrogen as a transportation fuel produced from thermal gasification of municipal solid waste: An examination of two integrated technologies [J]. International Journal of Hydrogen Energy, 2001, 26(11): 1209-1221.

[27] He M, Xiao B, Liu S, et al. Hydrogen-rich gas from catalytic steam gasification of municipal solid waste (MSW): Influence of steam to MSW ratios and weight hourly space velocity on gas production and composition [J]. International Journal of Hydrogen Energy, 2009, 34(5): 2174-2183.

[28] Wang J, Cheng G, You Y, et al. Hydrogen-rich gas production by steam gasification of municipal solid waste (MSW) using NiO supported on modified dolomite [J]. International Journal of Hydrogen Energy, 2012, 37(8): 6503-6510.

[29] Bellomare F, Rokni M. Integration of a municipal solid waste gasification plant with solid oxide fuel cell and gas turbine [J]. Renewable Energy, 2013, 55: 490-500.

[30] Arena U. Process and technological aspects of municipal solid waste gasification: A review [J]. Waste Management, 2012, 32(4): 625-639.

[31] Jiang H, Zhu X, Guo Q, et al. Gasification of rice husk in a fluidized-bed gasifier without inert additives [J]. Industrial & Engineering Chemistry Research, 2003, 42(23): 5745-5750.

[32] Toor S S, Rosendahl L, Rudolf A. Hydrothermal liquefaction of biomass: A review of subcritical water technologies [J]. Energy, 2011, 36(5): 2328-2342.

[33] Zhang L, Champagne P, Xu C. Bio-crude production from secondary pulp/paper-mill sludge and waste newspaper via co-liquefaction in hot-compressed

water [J]. Energy, 2011, 36(4): 2142-2150.

[34] Akhtar J, Amin N a S. A review on process conditions for optimum bio-oil yield in hydrothermal liquefaction of biomass [J]. Renewable and Sustainable Energy Reviews, 2011, 15(3): 1615-1624.

[35] Hwang I-H, Aoyama H, Matsuto T, et al. Recovery of solid fuel from municipal solid waste by hydrothermal treatment using subcritical water [J]. Waste Management, 2012, 32(3): 410-416.

[36] Bhaskar T, Matsui T, Kaneko J, et al. Novel calcium based sorbent (Ca-C) for the dehalogenation (Br, Cl) process during halogenated mixed plastic (PP/PE/PS/PVC and HIPS-Br) pyrolysis [J]. Green Chemistry, 2002, 4(4): 372-375.

[37] Liu W J, Li W W, Jiang H, et al. Fates of chemical elements in biomass during its pyrolysis [J]. Chemical Reviews, 2017, 117(9): 6367-6398.

[38] Huber G W, Iborra S, Corma A. Synthesis of transportation fuels from biomass: chemistry, catalysts, and engineering [J]. Chemical Reviews, 2006, 106(9): 4044-4098.

[39] Czernik S, Bridgwater A V. Overview of applications of biomass fast pyrolysis oil [J]. Energy & Fuels, 2004, 18(2): 590-598.

[40] Liu W J, Jiang H, Yu H Q. Development of biochar-based functional materials: Toward a austainable platform carbon material [J]. Chemical Reviews, 2015, 115(22): 12251-12285.

[41] Liu W J, Jiang H, Yu H Q. Emerging applications of biochar-based materials for energy storage and conversion [J]. Energy & Environmental Science, 2019, 12(6): 1751-1779.

[42] Borghei M, Lehtonen J, Liu L, et al. Advanced biomass-derived electrocatalysts for the oxygen reduction reaction [J]. Advanced Materials, 2017: 1703691.

[43] Woolf D, Amonette J E, Street-Perrott F A, et al. Sustainable biochar to mitigate global climate change [J]. Nat Commun, 2010, 1: 56.

[44] Bridgwater A V. Review of fast pyrolysis of biomass and product upgrading [J]. Biomass and Bioenergy, 2012, 38: 68-94.

[45] Hu X, Gholizadeh M. Biomass pyrolysis: A review of the process development and challenges from initial researches up to the commercialisation stage [J]. Journal of Energy Chemistry, 2019, 39:

109-143.

[46] Babu B V. Biomass pyrolysis: A state-of-the-art review [J]. Biofuels, Bioproducts and Biorefining, 2008, 2(5): 393-414.

[47] Dhyani V, Bhaskar T. A comprehensive review on the pyrolysis of lignocellulosic biomass [J]. Renewable Energy, 2018, 129: 695-716.

[48] Desisto W J, Hill N, Beis S H, et al. Fast pyrolysis of pine sawdust in a fluidized-bed reactor [J]. Energy & Fuels, 2010, 24(4): 2642-2651.

[49] Alonso D M, Wettstein S G, Dumesic J A. Bimetallic catalysts for upgrading of biomass to fuels and chemicals [J]. Chemical Society Reviews, 2012, 41: 8075-8098.

[50] Binod P, Sindhu R, Singhania R R, et al. Bioethanol production from rice straw: An overview [J]. Bioresource Technology, 2010, 101(13): 4767-4774.

[51] Dhepe P L, Fukuoka A. Cellulose conversion under heterogeneous catalysis [J]. ChemSusChem, 2008, 1(12): 969-975.

[52] Rinaldi R, Schüth F. Acid hydrolysis of cellulose as the entry point into biorefinery schemes [J]. ChemSusChem, 2009, 2(12): 1096-1107.

[53] Wyman C E, Dale B E, Elander R T, et al. Coordinated development of leading biomass pretreatment technologies [J]. Bioresource Technology, 2005, 96(18): 1959-1966.

[54] Mosier N, Wyman C, Dale B, et al. Features of promising technologies for pretreatment of lignocellulosic biomass [J]. Bioresource Technology, 2005, 96(6): 673-686.

[55] Bridgwater A V, Meier D, Radlein D. An overview of fast pyrolysis of biomass [J]. Organic Geochemistry, 1999, 30(12): 1479-1493.

[56] Wang S, Dai G, Yang H, et al. Lignocellulosic biomass pyrolysis mechanism: A state-of-the-art review [J]. Progress in Energy and Combustion Science, 2017, 62: 33-86.

[57] Liu H, Ahmad M S, Alhumade H, et al. A hybrid kinetic and optimization approach for biomass pyrolysis: The hybrid scheme of the isoconversional methods, DAEM, and a parallel-reaction mechanism [J]. Energy Conversion and Management, 2020, 208: 112531.

[58] Li X, Yin C, Knudsen Kær S, et al. A detailed pyrolysis model for a thermally large biomass particle [J]. Fuel, 2020, 278: 118397.

[59] Kalogiannis K G, Stefanidis S D, Karakoulia S A, et al. First pilot scale study of basic vs acidic catalysts in biomass pyrolysis: deoxygenation mechanisms and catalyst deactivation [J]. Applied Catalysis B: Environmental, 2018, 238: 346-357.

[60] Gogoi M, Konwar K, Bhuyan N, et al. Assessments of pyrolysis kinetics and mechanisms of biomass residues using thermogravimetry [J]. Bioresource Technology Reports, 2018, 4: 40-49.

[61] Dai G, Zhu Y, Yang J, et al. Mechanism study on the pyrolysis of the typical ether linkages in biomass [J]. Fuel, 2019, 249: 146-153.

[62] Chen Y, Liu B, Yang H, et al. Generalized two-dimensional correlation infrared spectroscopy to reveal the mechanisms of lignocellulosic biomass pyrolysis [J]. Proceedings of the Combustion Institute, 2019, 37(3): 3013-3021.

[63] Shen D K, Gu S, Bridgwater A V. Study on the pyrolytic behaviour of xylan-based hemicellulose using TG-FTIR and Py-GC-FTIR [J]. Journal of Analytical and Applied Pyrolysis, 2010, 87(2): 199-206.

[64] Peters B. Prediction of pyrolysis of pistachio shells based on its components hemicellulose, cellulose and lignin [J]. Fuel Processing Technology, 2011, 92(10): 1993-1998.

[65] Peng Y, Wu S. The structural and thermal characteristics of wheat straw hemicellulose [J]. Journal of Analytical and Applied Pyrolysis, 2010, 88(2): 134-139.

[66] Huang J, Liu C, Tong H, et al. Theoretical studies on pyrolysis mechanism of xylopyranose [J]. Computational and Theoretical Chemistry, 2012, 1001: 44-50.

[67] Patwardhan P R, Brown R C, Shanks B H. Product distribution from the fast pyrolysis of hemicellulose [J]. ChemSusChem, 2011, 4(5): 636-643.

[68] Sefain M Z, El-Kalyoubi S F, Shukry N. Thermal behavior of holo- and hemicellulose obtained from rice straw and bagasse [J]. Journal of Polymer Science: Polymer Chemistry Edition, 1985, 23(5): 1569-1577.

[69] Li S, Lyons-Hart J, Banyasz J, et al. Real-time evolved gas analysis by FTIR method: An experimental study of cellulose pyrolysis [J]. Fuel, 2001, 80(12): 1809-1817.

[70] Patwardhan P R, Dalluge D L, Shanks B H, et al. Distinguishing primary

and secondary reactions of cellulose pyrolysis [J]. Bioresource Technology, 2011, 102(8): 5265-5269.

[71] Zhang X, Yang W, Blasiak W. Thermal decomposition mechanism of levoglucosan during cellulose pyrolysis [J]. Journal of Analytical and Applied Pyrolysis, 2012, 96: 110-119.

[72] Banyasz J L, Li S, Lyons-Hart J, et al. Gas evolution and the mechanism of cellulose pyrolysis [J]. Fuel, 2001, 80(12): 1757-63.

[73] Ronsse F, Bai X, Prins W, et al. Secondary reactions of levoglucosan and char in the fast pyrolysis of cellulose [J]. Environmental Progress & Sustainable Energy, 2012, 31(2): 256-60.

[74] Mettler M S, Paulsen A D, Vlachos D G, et al. Pyrolytic conversion of cellulose to fuels: Levoglucosan deoxygenation via elimination and cyclization within molten biomass [J]. Energy & Environmental Science, 2012, 5(7): 7864-7868.

[75] Lin Y-C, Cho J, Tompsett G A, et al. Kinetics and mechanism of cellulose pyrolysis [J]. The Journal of Physical Chemistry C, 2009, 113 (46): 20097-20107.

[76] Luo Z Y, Wang S R, Liao Y F, et al. Mechanism study of cellulose rapid pyrolysis [J]. Industrial & Engineering Chemistry Research, 2004, 43(18): 5605-5610.

[77] Wooten J B, Seeman J I, Hajaligol M R. Observation and characterization of cellulose pyrolysis intermediates by ^{13}C CPMAS NMR: A new mechanistic model [J]. Energy & Fuels, 2003, 18(1): 1-15.

[78] Vinu R, Broadbelt L J. A mechanistic model of fast pyrolysis of glucose-based carbohydrates to predict bio-oil composition [J]. Energy & Environmental Science, 2012, 5(12): 9808-9826.

[79] Kosa M, Ben H, Theliander H, et al. Pyrolysis oils from CO_2 precipitated Kraft lignin [J]. Green Chemistry, 2011, 13(11): 3196.

[80] Chu S, Subrahmanyam A V, Huber G W. The pyrolysis chemistry of a [small beta]-O-4 type oligomeric lignin model compound [J]. Green Chemistry, 2012.

[81] Cho J, Chu S, Dauenhauer P J, et al. Kinetics and reaction chemistry for slow pyrolysis of enzymatic hydrolysis lignin and organosolv extracted lignin derived from maplewood [J]. Green Chemistry, 2012, 14(2): 428.

[82] Mu W, Ben H, Ragauskas A, et al. Lignin pyrolysis components and upgrading: Technology review [J]. Bioenerg Res, 2013: 1-22.

[83] Ben H, Ragauskas A J. NMR characterization of pyrolysis oils from Kraft lignin [J]. Energy & Fuels, 2011, 25(5): 2322-2332.

[84] Akhtar J, Saidina Amin N. A review on operating parameters for optimum liquid oil yield in biomass pyrolysis [J]. Renewable and Sustainable Energy Reviews, 2012, 16(7): 5101-5109.

[85] Sharifzadeh M, Sadeqzadeh M, Guo M, et al. The multi-scale challenges of biomass fast pyrolysis and bio-oil upgrading: Review of the state of art and future research directions [J]. Progress in Energy and Combustion Science, 2019, 71: 1-80.

[86] Pattiya A. Bio-oil production via fast pyrolysis of biomass residues from cassava plants in a fluidised-bed reactor [J]. Bioresource Technology, 2011, 102(2): 1959-1967.

[87] Shen D K, Gu S. The mechanism for thermal decomposition of cellulose and its main products [J]. Bioresource Technology, 2009, 100(24): 6496-6504.

[88] Uzun B B, Pütün A E, Pütün E. Fast pyrolysis of soybean cake: Product yields and compositions [J]. Bioresource Technology, 2006, 97(4): 569-576.

[89] Kan T, Strezov V, Evans T J. Lignocellulosic biomass pyrolysis: A review of product properties and effects of pyrolysis parameters [J]. Renewable and Sustainable Energy Reviews, 2016, 57: 1126-1140.

[90] Fu Q, Argyropoulos D S, Tilotta D C, et al. Understanding the pyrolysis of CCA-treated wood. Part II: Effect of phosphoric acid [J]. Journal of Analytical and Applied Pyrolysis, 2008, 82(1): 140-144.

[91] Sánchez C. Lignocellulosic residues: Biodegradation and bioconversion by fungi [J]. Biotechnology Advances, 2009, 27(2): 185-194.

[92] Zhang S, Yan Y, Li T, et al. Upgrading of liquid fuel from the pyrolysis of biomass [J]. Bioresource Technology, 2005, 96(5): 545-550.

[93] Horne P A, Williams P T. Influence of temperature on the products from the flash pyrolysis of biomass [J]. Fuel, 1996, 75(9): 1051-1059.

[94] Chan W-C R, Kelbon M, Krieger B B. Modelling and experimental verification of physical and chemical processes during pyrolysis of a large biomass particle [J]. Fuel, 1985, 64(11): 1505-1513.

[95] Strezov V, Moghtaderi B, Lucas J A. Thermal study of decomposition of selected biomass samples [J]. Journal of Thermal Analysis and Calorimetry, 2003, 72(3): 1041-1048.

[96] Debdoubi A, Amarti A E, Colacio E, et al. The effect of heating rate on yields and compositions of oil products from esparto pyrolysis [J]. International Journal of Energy Research, 2006, 30(15): 1243-1250.

[97] Tsai W, Lee M, Chang Y. Fast pyrolysis of rice husk: Product yields and compositions [J]. Bioresource Technology, 2007, 98(1): 22-28.

[98] Salehi E, Abedi J, Harding T. Bio-oil from sawdust: Pyrolysis of sawdust in a fixed-bed system [J]. Energy & Fuels, 2009, 23(7): 3767-3772.

[99] Sensöz S, Angın D. Pyrolysis of safflower (Charthamus tinctorius L.) seed press cake in a fixed-bed reactor. Part 2: Structural characterization of pyrolysis bio-oils [J]. Bioresource Technology, 2008, 99(13): 5498-5504.

[100] Angın D. Effect of pyrolysis temperature and heating rate on biochar obtained from pyrolysis of safflower seed press cake [J]. Bioresource Technology, 2013, 128: 593-597.

[101] Ozbay N, Pütün A E, Pütün E. Bio-oil production from rapid pyrolysis of cottonseed cake: Product yields and compositions [J]. International Journal of Energy Research, 2006, 30(7): 501-510.

[102] Aylón E, Fernández-Colino A, Navarro M V, et al. Waste tire pyrolysis: Comparison between fixed bed reactor and moving bed reactor [J]. Industrial & Engineering Chemistry Research, 2008, 47(12): 4029-4033.

[103] Onay O, Mete-Koçkar O. Fixed-bed pyrolysis of rapeseed (Brassica napus L.) [J]. Biomass and Bioenergy, 2004, 26(3): 289-299.

[104] Encinar J M, Beltrán F J, Bernalte A, et al. Pyrolysis of two agricultural residues: Olive and grape bagasse. Influence of particle size and temperature [J]. Biomass and Bioenergy, 1996, 11(5): 397-409.

[105] Olukcu N, Yanik J, Saglam M, et al. Liquefaction of beypazari oil shale by pyrolysis [J]. Journal of Analytical and Applied Pyrolysis, 2002, 64(1): 29-41.

[106] Bridgwater A V. Principles and practice of biomass fast pyrolysis processes for liquids [J]. Journal of Analytical and Applied Pyrolysis, 1999, 51(1/2): 3-22.

[107] Scott D S, Majerski P, Piskorz J, et al. A second look at fast pyrolysis of

biomass: The RTI process [J]. Journal of Analytical and Applied Pyrolysis, 1999, 51(1/2): 23-37.

[108] Acıkgoz C, Onay O, Kockar O M. Fast pyrolysis of linseed: Product yields and compositions [J]. Journal of Analytical and Applied Pyrolysis, 2004, 71(2): 417-429.

[109] El Harfi K, Mokhlisse A, Chanâa M B. Effect of water vapor on the pyrolysis of the Moroccan (Tarfaya) oil shale [J]. Journal of Analytical and Applied Pyrolysis, 1999, 48(2): 65-76.

[110] Gerçel H F. The effect of a sweeping gas flow rate on the fast pyrolysis of biomass [J]. Energy sources, 2002, 24(7): 633-642.

[111] Gani A, Naruse I. Effect of cellulose and lignin content on pyrolysis and combustion characteristics for several types of biomass [J]. Renewable Energy, 2007, 32(4): 649-661.

[112] Demirbas A. Effect of initial moisture content on the yields of oily products from pyrolysis of biomass [J]. Journal of Analytical and Applied Pyrolysis, 2004, 71(2): 803-815.

[113] Shen L, Zhang D-K. Low-temperature pyrolysis of sewage sludge and putrescible garbage for fuel oil production [J]. Fuel, 2005, 84(7/8): 809-815.

[114] Pattiya A, Suttibak S. Production of bio-oil via fast pyrolysis of agricultural residues from cassava plantations in a fluidised-bed reactor with a hot vapour filtration unit [J]. Journal of Analytical and Applied Pyrolysis, 2012, 95: 227-235.

[115] Chen T, Wu C, Liu R, et al. Effect of hot vapor filtration on the characterization of bio-oil from rice husks with fast pyrolysis in a fluidized-bed reactor [J]. Bioresource Technology, 2011, 102(10): 6178-6185.

[116] Lu Q, Li W Z, Zhu X F. Overview of fuel properties of biomass fast pyrolysis oils [J]. Energy Conversion and Management, 2009, 50(5): 1376-83.

[117] Javaid A, Ryan T, Berg G, et al. Removal of char particles from fast pyrolysis bio-oil by microfiltration [J]. Journal of Membrane Science, 2010, 363(1/2): 120-127.

[118] Teella A, Huber G W, Ford D M. Separation of acetic acid from the aqueous fraction of fast pyrolysis bio-oils using nanofiltration and reverse

osmosis membranes [J]. Journal of Membrane Science, 2011, 378(1/2): 495-502.

[119] Chiaramonti D, Bonini M, Fratini E, et al. Development of emulsions from biomass pyrolysis liquid and diesel and their use in engines. Part 2: Tests in diesel engines [J]. Biomass and Bioenergy, 2003, 25(1): 101-111.

[120] Chiaramonti D, Bonini M, Fratini E, et al. Development of emulsions from biomass pyrolysis liquid and diesel and their use in engines. Part 1: Emulsion production [J]. Biomass and Bioenergy, 2003, 25(1): 85-99.

[121] Guo Z, Wang S, Wang X. Stability mechanism investigation of emulsion fuels from biomass pyrolysis oil and diesel [J]. Energy, 2014, 66: 250-255.

[122] Diebold J P, Czernik S. Additives to lower and stabilize the viscosity of pyrolysis oils during storage [J]. Energy & Fuels, 1997, 11(5): 1081-1091.

[123] Deng L, Yan Z, Fu Y, et al. Green solvent for flash pyrolysis oil separation [J]. Energy & Fuels, 2009, 23(6): 3337-8.

[124] Zhao W, Zhang X, Huang J, et al. Hydrogenation of bio-oil via gas-liquid two-phase discharge reaction system [J]. Process Safety and Environmental Protection, 2018, 118: 167-177.

[125] Zhang X, Wang K, Chen J, et al. Mild hydrogenation of bio-oil and its derived phenolic monomers over Pt-Ni bimetal-based catalysts [J]. Applied Energy, 2020, 275: 115154.

[126] Xu Y, Li Y, Wang C, et al. In-situ hydrogenation of model compounds and raw bio-oil over Ni/CMK-3 catalyst [J]. Fuel Processing Technology, 2017, 161: 226-231.

[127] Shafaghat H, Tsang Y F, Jeon J-K, et al. In-situ hydrogenation of bio-oil/bio-oil phenolic compounds with secondary alcohols over a synthesized mesoporous Ni/CeO$_2$ catalyst [J]. Chemical Engineering Journal, 2020, 382: 122912.

[128] Mondal A K, Qin C, Ragauskas A J, et al. Conversion of Loblolly pine biomass residues to bio-oil in a two-step process: Fast pyrolysis in the presence of zeolite and catalytic hydrogenation [J]. Industrial Crops and Products, 2020, 148: 112318.

[129] Nava R, Pawelec B, Castaño P, et al. Upgrading of bio-liquids on different mesoporous silica-supported CoMo catalysts [J]. Applied Catalysis B: Environmental, 2009, 92(1/2): 154-67.

[130] Krár M, Kovács S, Kalló D, et al. Fuel purpose hydrotreating of sunflower oil on CoMo/Al$_2$O$_3$ catalyst [J]. Bioresource Technology, 2010, 101(23): 9287-9293.

[131] Srifa A, Faungnawakij K, Itthibenchapong V, et al. Production of bio-hydrogenated diesel by catalytic hydrotreating of palm oil over NiMoS$_2$/γ-Al$_2$O$_3$ catalyst [J]. Bioresource Technology, 2014, 158: 81-90.

[132] Bezergianni S, Dimitriadis A, Meletidis G. Effectiveness of CoMo and NiMo catalysts on co-hydroprocessing of heavy atmospheric gas oil-waste cooking oil mixtures [J]. Fuel, 2014, 125: 129-36.

[133] Tang Y, Miao S, Pham H N, et al. Enhancement of Pt catalytic activity in the hydrogenation of aldehydes [J]. Applied Catalysis A: General, 2011, 406(1/2): 81-88.

[134] Tang Y, Miao S, Shanks B H, et al. Bifunctional mesoporous organic-inorganic hybrid silica for combined one-step hydrogenation/esterification [J]. Applied Catalysis A: General, 2010, 375(2): 310-317.

[135] Yu W, Tang Y, Mo L, et al. Bifunctional Pd/Al-SBA-15 catalyzed one-step hydrogenation-esterification of furfural and acetic acid: A model reaction for catalytic upgrading of bio-oil [J]. Catalysis Communications, 2011, 13(1): 35-39.

[136] Wan H, Vitter A, Chaudhari R V, et al. Kinetic investigations of unusual solvent effects during Ru/C catalyzed hydrogenation of model oxygenates [J]. Journal of Catalysis, 2014, 309: 174-184.

[137] Xiong W-M, Fu Y, Zeng F-X, et al. An in situ reduction approach for bio-oil hydroprocessing [J]. Fuel Processing Technology, 2011, 92(8): 1599-605.

[138] Ying X, Tiejun W, Longlong M, et al. Upgrading of fast pyrolysis liquid fuel from biomass over Ru/γ-Al$_2$O$_3$ catalyst [J]. Energy Conversion and Management, 2012, 55: 172-177.

[139] Yu W, Tang Y, Mo L, et al. One-step hydrogenation-esterification of furfural and acetic acid over bifunctional Pd catalysts for bio-oil upgrading [J]. Bioresource Technology, 2011, 102(17): 8241-8246.

[140] Busetto L, Fabbri D, Mazzoni R, et al. Application of the Shvo catalyst in homogeneous hydrogenation of bio-oil obtained from pyrolysis of white poplar: New mild upgrading conditions [J]. Fuel, 2011, 90(3):

1197-1207.

[141] Elkasabi Y, Mullen C A, Pighinelli A L M T, et al. Hydrodeoxygenation of fast-pyrolysis bio-oils from various feedstocks using carbon-supported catalysts [J]. Fuel Processing Technology, 2014, 123: 11-18.

[142] Zhong W C, Guo Q J, Wang X Y, et al. Catalytic hydroprocessing of fast pyrolysis bio-oil from Chlorella [J]. Journal of Fuel Chemistry and Technology, 2013, 41(5): 571-578.

[143] Ben H, Mu W, Deng Y, et al. Production of renewable gasoline from aqueous phase hydrogenation of lignin pyrolysis oil [J]. Fuel, 2013, 103: 1148-1153.

[144] Mahfud F H, Ghijsen F, Heeres H J. Hydrogenation of fast pyrolyis oil and model compounds in a two-phase aqueous organic system using homogeneous ruthenium catalysts [J]. Journal of Molecular Catalysis A: Chemical, 2007, 264(1/2): 227-236.

[145] Bykova M V, Ermakov D Y, Khromova S A, et al. Stabilized Ni-based catalysts for bio-oil hydrotreatment: Reactivity studies using guaiacol [J]. Catalysis Today, 2014, 220-222: 21-31.

[146] Zhang X, Wang T, Ma L, et al. Hydrotreatment of bio-oil over Ni-based catalyst [J]. Bioresource Technology, 2013, 127: 306-311.

[147] Kim T-S, Oh S, Kim J-Y, et al. Study on the hydrodeoxygenative upgrading of crude bio-oil produced from woody biomass by fast pyrolysis [J]. Energy, 2014, 68: 437-43.

[148] Xu X, Zhang C, Liu Y, et al. Two-step catalytic hydrodeoxygenation of fast pyrolysis oil to hydrocarbon liquid fuels [J]. Chemosphere, 2013, 93 (4): 652-660.

[149] Kaewpengkrow P, Atong D, Sricharoenchaikul V. Effect of Pd, Ru, Ni and ceramic supports on selective deoxygenation and hydrogenation of fast pyrolysis Jatropha residue vapors [J]. Renewable Energy, 2014, 65: 92-101.

[150] Echeandia S, Pawelec B, Barrio V L, et al. Enhancement of phenol hydrodeoxygenation over Pd catalysts supported on mixed HY zeolite and Al_2O_3. An approach to O-removal from bio-oils [J]. Fuel, 2014, 117, Part B: 1061-1073.

[151] Deng L, Li J, Lai D-M, et al. Catalytic conversion of biomass-derived

carbohydrates into γ-valerolactone without using an external H_2 supply [J]. Angewandte Chemie International Edition, 2009, 48(35): 6529-6532.

[152] Chen X, Che Q, Li S, et al. Recent developments in lignocellulosic biomass catalytic fast pyrolysis: Strategies for the optimization of bio-oil quality and yield [J]. Fuel Processing Technology, 2019, 196: 106180.

[153] Ochoa A, Bilbao J, Gayubo A G, et al. Coke formation and deactivation during catalytic reforming of biomass and waste pyrolysis products: A review [J]. Renewable and Sustainable Energy Reviews, 2020, 119: 109600.

[154] Rahman M M, Liu R, Cai J. Catalytic fast pyrolysis of biomass over zeolites for high quality bio-oil: A review [J]. Fuel Processing Technology, 2018, 180: 32-46.

[155] Nishu, Liu R, Rahman M M, et al. A review on the catalytic pyrolysis of biomass for the bio-oil production with ZSM-5: Focus on structure [J]. Fuel Processing Technology, 2020, 199: 106301.

[156] Bond J Q, Upadhye A A, Olcay H, et al. Production of renewable jet fuel range alkanes and commodity chemicals from integrated catalytic processing of biomass [J]. Energy & Environmental Science, 2014, 7(4): 1500-1523.

[157] Carlson T R, Cheng Y-T, Jae J, et al. Production of green aromatics and olefins by catalytic fast pyrolysis of wood sawdust [J]. Energy & Environmental Science, 2011, 4(1): 145-161.

[158] Iliopoulou E F, Antonakou E V, Karakoulia S A, et al. Catalytic conversion of biomass pyrolysis products by mesoporous materials: Effect of steam stability and acidity of Al-MCM-41 catalysts [J]. Chemical Engineering Journal, 2007, 134(1/2/3): 51-57.

[159] Zhang H, Carlson T R, Xiao R, et al. Catalytic fast pyrolysis of wood and alcohol mixtures in a fluidized bed reactor [J]. Green Chemistry, 2012, 14(1): 98.

[160] Veses A, Aznar M, Martínez I, et al. Catalytic pyrolysis of wood biomass in an auger reactor using calcium-based catalysts [J]. Bioresource Technology, 2014, 162: 250-258.

[161] Naqvi S R, Uemura Y, Yusup S B. Catalytic pyrolysis of paddy husk in a drop type pyrolyzer for bio-oil production: The role of temperature and catalyst [J]. Journal of Analytical and Applied Pyrolysis, 2014, 106:

57-62.

[162] Zheng Y, Chen D, Zhu X. Aromatic hydrocarbon production by the online catalytic cracking of lignin fast pyrolysis vapors using $Mo_2N/\gamma\text{-}Al_2O_3$[J]. Journal of Analytical and Applied Pyrolysis, 2013, 104: 514-20.

[163] Shadangi K P, Mohanty K. Thermal and catalytic pyrolysis of Karanja seed to produce liquid fuel [J]. Fuel, 2014, 115: 434-442.

[164] Zheng A, Zhao Z, Chang S, et al. Effect of crystal size of ZSM-5 on the aromatic yield and selectivity from catalytic fast pyrolysis of biomass [J]. Journal of Molecular Catalysis A: Chemical, 2014, 383-384: 23-30.

[165] Mante O D, Agblevor F A, Mcclung R. A study on catalytic pyrolysis of biomass with Y-zeolite based FCC catalyst using response surface methodology [J]. Fuel, 2013, 108: 451-464.

[166] Mullen C A, Boateng A A, Mihalcik D J, et al. Catalytic fast pyrolysis of white oak wood in a bubbling fluidized bed [J]. Energy & Fuels, 2011, 25(11): 5444-5451.

[167] Zhang H, Nie J, Xiao R, et al. Catalytic co-pyrolysis of biomass and different plastics (polyethylene, polypropylene, and polystyrene) to improve hydrocarbon yield in a fluidized-bed reactor [J]. Energy & Fuels, 2014, 28(3): 1940-1947.

[168] Lu Q, Tang Z, Zhang Y, et al. Catalytic upgrading of biomass fast pyrolysis vapors with Pd/SBA-15 catalysts [J]. Industrial & Engineering Chemistry Research, 2010, 49(6): 2573-2580.

[169] Thangalazhy-Gopakumar S, Adhikari S, Gupta R B. Catalytic pyrolysis of biomass over H + ZSM-5 under hydrogen pressure [J]. Energy & Fuels, 2012, 26(8): 5300-5306.

[170] Srinivasan V, Adhikari S, Chattanathan S A, et al. Catalytic pyrolysis of torrefied biomass for hydrocarbons production [J]. Energy & Fuels, 2012, 26(12): 7347-7353.

[171] Shabaker J, Huber G, Dumesic J. Aqueous-phase reforming of oxygenated hydrocarbons over Sn-modified Ni catalysts [J]. Journal of Catalysis, 2004, 222(1): 180-191.

[172] Davda R, Shabaker J, Huber G, et al. A review of catalytic issues and process conditions for renewable hydrogen and alkanes by aqueous-phase reforming of oxygenated hydrocarbons over supported metal catalysts [J].

Applied Catalysis B: Environmental, 2005, 56(1): 171-186.

[173] Huber G W, Chheda J, Barrett C, et al. Production of liquid alkanes by aqueous-phase processing of biomass-derived carbohydrates [J]. Science, 2005, 308: 1446-2079.

[174] Chheda J N, Huber G W, Dumesic J A. Liquid-phase catalytic processing of biomass-derived oxygenated hydrocarbons to fuels and chemicals [J]. Angewandte Chemie International Edition, 2007, 46(38): 7164-7183.

[175] Borg Ø, Hammer N, Enger B C, et al. Effect of biomass-derived synthesis gas impurity elements on cobalt Fischer-Tropsch catalyst performance including in situ sulphur and nitrogen addition [J]. Journal of Catalysis, 2011, 279(1): 163-173.

[176] Zwart R W, Boerrigter H. High efficiency co-production of Synthetic Natural Gas (SNG) and Fischer-Tropsch (FT) transportation fuels from biomass [J]. Energy & Fuels, 2005, 19(2): 591-597.

[177] Wright M M, Brown R C, Boateng A A. Distributed processing of biomass to bio-oil for subsequent production of Fischer-Tropsch liquids [J]. Biofuels, Bioproducts and Biorefining, 2008, 2(3): 229-238.

[178] Tijmensen M J, Faaij A P, Hamelinck C N, et al. Exploration of the possibilities for production of Fischer-Tropsch liquids and power via biomass gasification [J]. Biomass and Bioenergy, 2002, 23(2): 129-152.

[179] Hu B, Wang K, Wu L, et al. Engineering carbon materials from the hydrothermal carbonization process of biomass [J]. Advanced Materials, 2010, 22(7): 813-828.

[180] Titirici M-M, Thomas A, Antonietti M. Back in the black: Hydrothermal carbonization of plant material as an efficient chemical process to treat the CO_2 problem? [J]. New Journal of Chemistry, 2007, 31(6): 787-789.

[181] Hu B, Yu S-H, Wang K, et al. Functional carbonaceous materials from hydrothermal carbonization of biomass: An effective chemical process [J]. Dalton Transactions, 2008, (40): 5414-5423.

[182] Titirici M-M, Antonietti M, Baccile N. Hydrothermal carbon from biomass: A comparison of the local structure from poly-to monosaccharides and pentoses/hexoses [J]. Green Chemistry, 2008, 10(11): 1204-1212.

[183] Xu Y J, Weinberg G, Liu X, et al. Nanoarchitecturing of activated carbon: Facile strategy for chemical functionalization of the surface of

activated carbon [J]. Advanced Functional Materials, 2008, 18(22): 3613-3619.

[184] Titirici M-M, White R J, Falco C, et al. Black perspectives for a green future: Hydrothermal carbons for environment protection and energy storage [J]. Energy & Environmental Science, 2012, 5(5): 6796-6822.

[185] Mohamad Nor N, Lau L C, Lee K T, et al. Synthesis of activated carbon from lignocellulosic biomass and its applications in air pollution control: A review [J]. Journal of Environmental Chemical Engineering, 2013, 1(4): 658-666.

[186] Ioannidou O, Zabaniotou A. Agricultural residues as precursors for activated carbon production: A review [J]. Renewable and Sustainable Energy Reviews, 2007, 11(9): 1966-2005.

[187] Carrott P, Ribeiro Carrott M. Lignin-from natural adsorbent to activated carbon: A review [J]. Bioresource Technology, 2007, 98(12): 2301-2312.

[188] Sun K. Preparation and characterization of activated carbon from rubber-seed shell by physical activation with steam [J]. Biomass and Bioenergy, 2010, 34(4): 539-544.

[189] Nuithitikul K, Srikhun S, Hirunpraditkoon S. Influences of pyrolysis condition and acid treatment on properties of durian peel-based activated carbon [J]. Bioresource Technology, 2010, 101(1): 426-429.

[190] Sumathi S, Bhatia S, Lee K, et al. Cerium impregnated palm shell activated carbon (Ce/PSAC) sorbent for simultaneous removal of SO_2 and NO-process study [J]. Chemical Engineering Journal, 2010, 162(1): 51-57.

[191] Nowicki P, Wachowska H, Pietrzak R. Active carbons prepared by chemical activation of plum stones and their application in removal of NO_2 [J]. Journal of Hazardous Materials, 2010, 181(1): 1088-1094.

[192] Nowicki P, Pietrzak R, Wachowska H. Sorption properties of active carbons obtained from walnut shells by chemical and physical activation [J]. Catalysis Today, 2010, 150(1): 107-114.

[193] ayan E. Ultrasound-assisted preparation of activated carbon from alkaline impregnated hazelnut shell: An optimization study on removal of Cu^{2+} from aqueous solution [J]. Chemical Engineering Journal, 2006, 115(3): 213-218.

[194] Demirbas A. Agricultural based activated carbons for the removal of dyes from aqueous solutions: A review [J]. Journal of Hazardous Materials, 2009, 167(1/2/3): 1-9.

[195] Arancon R A, Barros Jr H R, Balu A M, et al. Valorisation of corncob residues to functionalised porous carbonaceous materials for the simultaneous esterification/transesterification of waste oils [J]. Green Chemistry, 2011, 13(11): 3162.

[196] Balakrishnan M, Batra V S, Hargreaves J S J, et al. Waste materials-catalytic opportunities: An overview of the application of large scale waste materials as resources for catalytic applications [J]. Green Chemistry, 2011, 13(1): 16.

[197] Biswal M, Banerjee A, Deo M, et al. From dead leaves to high energy density supercapacitors [J]. Energy & Environmental Science, 2013, 6(4): 1249-1259.

[198] Guo Y, Qi J, Jiang Y, et al. Performance of electrical double layer capacitors with porous carbons derived from rice husk [J]. Materials Chemistry and Physics, 2003, 80(3): 704-709.

[199] Wu F C, Tseng R L, Hu C C, et al. Physical and electrochemical characterization of activated carbons prepared from firwoods for supercapacitors [J]. Journal of Power Sources, 2004, 138(1): 351-359.

[200] Wu F C, Tseng R L, Hu C C, et al. Effects of pore structure and electrolyte on the capacitive characteristics of steam-and KOH-activated carbons for supercapacitors [J]. Journal of Power Sources, 2005, 144(1): 302-309.

[201] Kim Y J, Lee B J, Suezaki H, et al. Preparation and characterization of bamboo-based activated carbons as electrode materials for electric double layer capacitors [J]. Carbon, 2006, 44(8): 1592-1595.

[202] Subramanian V, Luo C, Stephan A, et al. Supercapacitors from activated carbon derived from banana fibers [J]. The Journal of Physical Chemistry C, 2007, 111(20): 7527-7531.

[203] Balathanigaimani M, Shim W G, Lee M J, et al. Highly porous electrodes from novel corn grains-based activated carbons for electrical double layer capacitors [J]. Electrochemistry Communications, 2008, 10(6): 868-871.

[204] Rufford T E, Hulicova-Jurcakova D, Zhu Z, et al. Nanoporous carbon

electrode from waste coffee beans for high performance supercapacitors [J]. Electrochemistry Communications, 2008, 10(10): 1594-1597.

[205] Kalpana D, Cho S, Lee S, et al. Recycled waste paper: A new source of raw material for electric double-layer capacitors [J]. Journal of Power Sources, 2009, 190(2): 587-591.

[206] Raymundo-Piñero E, Cadek M, Beguin F. Tuning carbon materials for supercapacitors by direct pyrolysis of seaweeds [J]. Advanced Functional Materials, 2009, 19(7): 1032-1039.

[207] Rufford T E, Hulicova-Jurcakova D, Khosla K, et al. Microstructure and electrochemical double-layer capacitance of carbon electrodes prepared by zinc chloride activation of sugar cane bagasse [J]. Journal of Power Sources, 2010, 195(3): 912-918.

[208] Ismanto A E, Wang S, Soetaredjo F E, et al. Preparation of capacitor's electrode from cassava peel waste [J]. Bioresource Technology, 2010, 101 (10): 3534-3540.

[209] Li X, Xing W, Zhuo S, et al. Preparation of capacitor's electrode from sunflower seed shell [J]. Bioresource Technology, 2011, 102 (2): 1118-1123.

[210] Ding J, Zhong J, Yang Y, et al. Occurrence and exposure to polycyclic aromatic hydrocarbons and their derivatives in a rural Chinese home through biomass fuelled cooking [J]. Environmental Pollution, 2012, 169: 160-166.

[211] Shen G, Tao S, Wei S, et al. Emissions of parent, nitro, and oxygenated polycyclic aromatic hydrocarbons from residential wood combustion in rural China [J]. Environmental Science & Technology, 2012, 46 (15): 8123-8130.

[212] Fagernäs L, Kuoppala E, Simell P. Polycyclic aromatic hydrocarbons in birch wood slow pyrolysis products [J]. Energy & Fuels, 2012, 26(11): 6960-6970.

[213] Zhou H, Wu C, Onwudili J A, et al. Polycyclic aromatic hydrocarbons (PAH) formation from the pyrolysis of different municipal solid waste fractions [J]. Waste Management, 2015, 36: 136-146.

[214] Hale S E, Lehmann J, Rutherford D, et al. Quantifying the total and bioavailable polycyclic aromatic hydrocarbons and dioxins in biochars [J].

Environmental Science & Technology, 2012, 46(5): 2830-2838.

[215] Fernandes M B, Brooks P. Characterization of carbonaceous combustion residues. Ⅱ: Nonpolar organic compounds [J]. Chemosphere, 2003, 53(5): 447-458.

[216] Nakajima D, Nagame S, Kuramochi H, et al. Polycyclic aromatic hydrocarbon generation behavior in the process of carbonization of wood [J]. Bulletin of Environmental Contamination and Toxicology, 2007, 79(2): 221-225.

[217] Brown R A, Kercher A K, Nguyen T H, et al. Production and characterization of synthetic wood chars for use as surrogates for natural sorbents [J]. Organic Geochemistry, 2006, 37(3): 321-333.

[218] Zhurinsh A, Zandersons J, Dobele G. Slow pyrolysis studies for utilization of impregnated waste timber materials [J]. Journal of Analytical and Applied Pyrolysis, 2005, 74(1): 439-444.

[219] Singh B, Singh B P, Cowie A L. Characterisation and evaluation of biochars for their application as a soil amendment [J]. Soil Research, 2010, 48(7): 516-525.

[220] Jonker M T, Koelmans A A. Extraction of polycyclic aromatic hydrocarbons from soot and sediment: Solvent evaluation and implications for sorption mechanism [J]. Environmental Science & Technology, 2002, 36(19): 4107-4113.

[221] Salt D E, Blaylock M, Kumar N P, et al. Phytoremediation: A novel strategy for the removal of toxic metals from the environment using plants [J]. Nature Biotechnology, 1995, 13(5): 468-474.

[222] Padmavathiamma P K, Li L Y. Phytoremediation technology: Hyper-accumulation metals in plants [J]. Water, Air, and Soil Pollution, 2007, 184(12/3/4): 105-126.

[223] Mcgrath S, Zhao F, Lombi E. Plant and rhizosphere processes involved in phytoremediation of metal-contaminated soils [J]. Plant and Soil, 2001, 232(1/2): 207-214.

[224] Huang J W, Blaylock M J, Kapulnik Y, et al. Phytoremediation of uranium-contaminated soils: Role of organic acids in triggering uranium hyperaccumulation in plants [J]. Environmental Science & Technology, 1998, 32(13): 2004-2008.

[225] Rugh C L, Senecoff J F, Meagher R B, et al. Development of transgenic yellow poplar for mercury phytoremediation [J]. Nature Biotechnology, 1998, 16(10): 925-928.

[226] Stals M, Thijssen E, Vangronsveld J, et al. Flash pyrolysis of heavy metal contaminated biomass from phytoremediation: Influence of temperature, entrained flow and wood/leaves blended pyrolysis on the behaviour of heavy metals [J]. Journal of Analytical and Applied Pyrolysis, 2010, 87(1): 1-7.

[227] Stals M, Carleer R, Reggers G, et al. Flash pyrolysis of heavy metal contaminated hardwoods from phytoremediation: Characterisation of biomass, pyrolysis oil and char/ash fraction [J]. Journal of Analytical and Applied Pyrolysis, 2010, 89(1): 22-29.

[228] Lievens C, Carleer R, Cornelissen T, et al. Fast pyrolysis of heavy metal contaminated willow: Influence of the plant part [J]. Fuel, 2009, 88(8): 1417-1425.

[229] Lievens C, Yperman J, Cornelissen T, et al. Study of the potential valorisation of heavy metal contaminated biomass via phytoremediation by fast pyrolysis. Part II: Characterisation of the liquid and gaseous fraction as a function of the temperature [J]. Fuel, 2008, 87(10/11): 1906-1916.

[230] Lievens C, Yperman J, Vangronsveld J, et al. Study of the potential valorisation of heavy metal contaminated biomass via phytoremediation by fast pyrolysis. Part I: Influence of temperature, biomass species and solid heat carrier on the behaviour of heavy metals [J]. Fuel, 2008, 87(10/11): 1894-1905.

[231] Koppolu L, Agblevor F A, Clements L D. Pyrolysis as a technique for separating heavy metals from hyperaccumulators. Part II: Lab-scale pyrolysis of synthetic hyperaccumulator biomass [J]. Biomass and Bioenergy, 2003, 25(6): 651-663.

[232] Koppolu L, Prasad R, Davis Clements L. Pyrolysis as a technique for separating heavy metals from hyperaccumulators. Part III: Pilot-scale pyrolysis of synthetic hyperaccumulator biomass [J]. Biomass and Bioenergy, 2004, 26(5): 463-472.

[233] Ren Q, Zhao C. NO_x and N_2O precursors from biomass pyrolysis: Nitrogen transformation from amino acid [J]. Environmental Science & Technology,

2012, 46(7): 4236-4240.

[234] Ren Q, Zhao C. NO_x and N_2O precursors from biomass pyrolysis: Role of cellulose, hemicellulose and lignin [J]. Environmental Science & Technology, 2013, 47(15): 8955-8961.

[235] Becidan M, Skreiberg O, Hustad J E. NO_x and N_2O precursors (NH_3 and HCN) in pyrolysis of biomass residues [J]. Energy & Fuels, 2007, 21(2): 1173-1180.

[236] Chang L, Xie Z, Xie K C, et al. Formation of NO_x precursors during the pyrolysis of coal and biomass. Part Ⅳ: Effects of gas atmosphere on the formation of NH_3 and HCN [J]. Fuel, 2003, 82(10): 1159-1166.

[237] Li C Z, Tan L L. Formation of NO_x and SO_x precursors during the pyrolysis of coal and biomass. Part Ⅲ: Further discussion on the formation of HCN and NH_3 during pyrolysis [J]. Fuel, 2000, 79(15): 1899-1906.

[238] Tan L L, Li C Z. Formation of NO_x and SO_x precursors during the pyrolysis of coal and biomass. Part Ⅰ: Effects of reactor configuration on the determined yields of HCN and NH_3 during pyrolysis [J]. Fuel, 2000, 79(15): 1883-1889.

[239] Tian F J, Wu H, Yu J L, et al. Formation of NO_x precursors during the pyrolysis of coal and biomass. Part Ⅷ: Effects of pressure on the formation of NH_3 and HCN during the pyrolysis and gasification of Victorian brown coal in steam [J]. Fuel, 2005, 84(16): 2102-2108.

第 2 章

生物质热解过程中污染物的迁移转化过程及其机理解析

2.1
生物质表面化学修饰强化污染物去除及其机理研究

2.1.1
概述

水环境的重金属污染是困扰着人类的重大环境问题之一[1-2]。目前重金属污染处理方法包括沉降法、电化学方法、膜分离法以及植物修复法等[3-8]。这些方法都可以有效去除重金属，但是也存在着很多问题，如成本过高、耗时长以及产生大量的有毒有害的副产物等[9]。使用废弃生物质吸附水体中的重金属操作简单、成本低廉，且不会产生大量的毒害副产物[10]。大量文献报道了多种生物质都具有很强的重金属富集能力和去除能力[11-18]。

一般而言，生物质不具备多孔性的特点，其对重金属的吸附固定主要是由于它表面骨架上的稳定不可溶的功能基团，例如羧基、氨基、羟基等对重金属的化学作用[19-20]。然而大多数的生物质表面的功能基团有限，导致其对重金属的吸附能力不高[21]。通过表面化学修饰来丰富生物质表面可用于重金属吸附的功能基团，是提高其重金属吸附能力的有效方法之一[22-23]。乙二胺四乙酸（EDTA）是一个含有丰富羧基和氨基功能基团的化学试剂，可以作为生物质的表面化学修饰试剂，赋予被修饰生物质很强的金属螯合能力和离子交换能力[24]。Nagib等采用EDTA的酸酐对壳聚糖进行表面化学修饰，发现其能显著增强重金属镍的吸附能力[25]。Yu等采用类似的方法对面包酵母生物质进行修饰，结果表明，经过修饰的生物质对铅和铜的吸附能力大大增加[26]。然而在修饰过程中，他们使用的EDTA酸酐的两个活性酸酐基团具有同样的化学活性，都可以和生物质上的氨基或者羟基反应，从而减少了可被用于重金属吸附的羧基数量。而且，生物质上不同功能基团对重金属吸附的影响，及环境条件对吸附影响的机制尚不清楚。

在本章的研究工作中，我们分析了生物质的表面结构，据此提出了一种对生物质表面进行 EDTA 修饰的化学方法,即先用化学当量的二氯亚砜($SOCl_2$)对 EDTA 中的一个羧基进行化学活化,形成 EDTA 酰氯化合物或者单酸酐化合物,再由其对生物质进行反应,这样就可以保证 1 个 EDTA 分子中只有 1 个羧基参与修饰反应,而剩下的 3 个羧基可以用作对重金属吸附的功能基团。我们选择蒲草对初始生物质进行修饰,蒲草是一种广泛存在的一年生湿地植物,其成本低廉,性质稳定,并且含有大量的氨基酸和多糖类化合物,方便随后的修饰过程。本章研究的主要内容是通过 $SOCl_2$ 活化的 EDTA 对蒲草生物质进行表面化学修饰,并阐述其对重金属铅的吸附增强能力和可能的机理。

2.1.2
生物质的表面化学修饰过程及其机理

蒲草生物质随机采集自合肥市内的一块湿地中,采集过后,将生物质先用水清洗数次,然后切成小段在 105 ℃ 条件下干燥过夜,干燥过的生物质再用高速粉碎机粉碎,过筛收集粒径在 60~100 目的颗粒,然后在索氏提取器中用丙酮提取 4 h 除去生物质表面可溶性有机物,剩余的生物质在 80 ℃ 下干燥 12 h,然后储存在干燥器中备用(称为初始蒲草生物质,original *Typha angustifolia* biomass,OTAB)。将 $SOCl_2$ 用作活化试剂,二甲亚砜(DMSO)用作反应过程中的溶剂,重金属铅溶液通过溶解一定量的硝酸铅于纯水中来配置。实验中所用的化学试剂均为分析纯。

OTAB 化学修饰过程需要在无水的条件下进行,使用 DMSO 作为溶剂(使用之前,先用分子筛干燥过夜,除去其中的水分)。将 11.68 g(0.04 mol)的 EDTA 和 100 mL 的无水 DMSO 混合在一个 250 mL 的三口烧瓶中,剧烈搅拌使其溶解。随后将 4.76 g(0.04 mol)的 $SOCl_2$ 在剧烈搅拌的情况下通过一个恒压滴液漏斗缓慢加入三口烧瓶中。当 $SOCl_2$ 滴加完毕后,将 4.00 g 的 OTAB 迅速加入至三口烧瓶,在常温下继续搅拌反应 2 h,然后就可以获得化学修饰的蒲草生物质(chemically modified *Typha angustifolia* biomass,CMTAB)。将所得的混合物进行抽滤得到固体残渣,再依次使用 DMSO、丙酮、纯水、碳酸钠溶液($NaCO_3$,0.1 mol·L^{-1})、纯水和丙酮清洗,最后,获得的 CMTAB 在 80 ℃ 下干燥去除残留的丙酮,然后储存在干燥器中备用。

蒲草生物质的表面化学修饰过程如图 2.1 所示。修饰过程主要可以分成两个阶段,即 EDTA 的羧基活化阶段和生物质的化学修饰阶段。在羧基活化阶段,EDTA 和二氯亚砜反应(这是一个常用的羧基活化试剂),其中的羧基转化成高度活性的酰氯基团。为了使 EDTA 的四个羧基基团只有一个被酰氯化,在活化过程中,必须保证 EDTA 和二氯亚砜的摩尔比为 1∶1。在生物质的化学修饰阶段,EDTA 一酰氯化合物一方面可以直接和蒲草生物质表面大量存在的氨基或者羟基反应,从而接枝到生物质的表面;另一方面,由于其极高的反应活性,也可以发生分子内的反应,生成 EDTA 一酸酐化合物,再和生物质表面的氨基或羟基反应而接枝到生物质骨架结构上。和常用的 EDTA 二酸酐化合物(这是在文献中报道最多的 EDTA 修饰方法)相比,$SOCl_2$ 活化的 EDTA 修饰生物质的方法不仅可以保证只有一个 EDTA 羧基参与反应,从而尽可能地增加可用于重金属吸附的羧基量,而且该方法的反应也具有很大的优势(常温下反应 2 h 即可,文献报道需要在常温下反应 24 h,或者 60 ℃下反应 4 h)[25-26]。

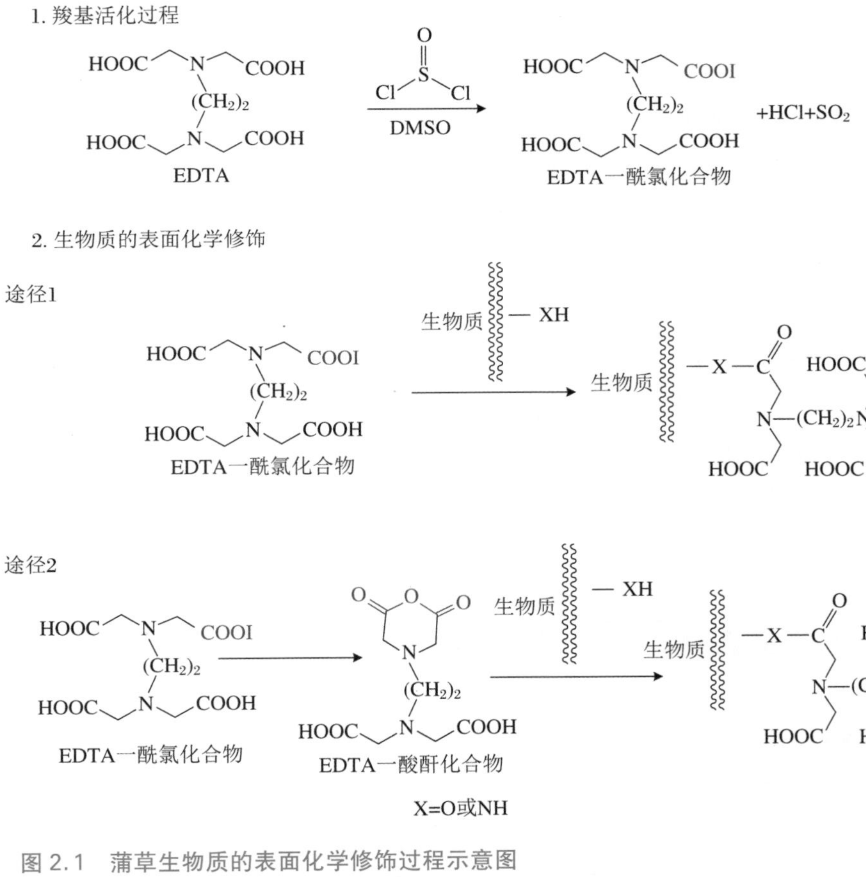

1. 羧基活化过程

2. 生物质的表面化学修饰

途径1

途径2

X=O 或 NH

图 2.1　蒲草生物质的表面化学修饰过程示意图

2.1.3
表面化学修饰生物质的结构和组成特征分析

OTAB 和 CMTAB 的总氮含量通过元素分析仪(VARIO EL Ⅲ,Elementar Inc.,Germany)进行测定,羧基基团的含量通过文献报道的反滴定法进行测定[27]:首先,准确称量的 1.0 g OTAB 或 CMTAB 样品和 100 mL 去离子水混合,然后调节溶液 pH 至 2.0,并持续搅拌 120 min,保证 COONa(在碳酸钠溶液洗涤过程中生成)全部转化为 COOH 基团。而后抽滤得到固体,将该固体干燥至质量不再改变。再准确称取 0.100 g 干燥固体,和 100 mL 的碳酸氢钠标准溶液($0.01\ mol\cdot L^{-1}$)混合,在氩气氛围下持续搅拌 120 min,随后立即过滤获得滤液。将所得滤液分成三份,分别用盐酸标准溶液($0.01\ mol\cdot L^{-1}$)进行滴定至中性,滴定过程中的 pH 变化通过 pH 计监测。通过消耗盐酸溶液的体积来计算羧酸基团的含量,其计算公式如下所示:

$$C_{COOH} = \frac{C_{NaHCO_3} \times V_{NaHCO_3} - C_{HCl} \times V_{HCl}}{W} \tag{2.1}$$

其中,C_{NaHCO_3},C_{HCl} 分别为碳酸氢钠和盐酸溶液的准确浓度,V_{NaHCO_3},V_{HCl} 分别为消耗碳酸氢钠和盐酸溶液的体积,W 为测定样品的质量。

红外光谱用来分析生物质化学修饰前后表面功能基团的变化,其测定过程为:一定量(约 10 mg)的样品与光谱纯的溴化钾晶体混合,研磨均匀之后压成片放入红外光谱仪(EQUIVOX55 IR spectroscopy,Bruker,Germany)进行测定。红外扫描范围是 4000～400 cm^{-1},扫描精度为 2 cm^{-1},每个样品扫描 16 次。修饰过程的形貌变化通过扫描电子显微镜(SEM,Sirion 200,FEI Electron Optics Company,USA)进行观察。生物质样品的表面结构和孔隙分布通过氮气吸脱附动力学进行分析,测定过程为:在 77 K 温度下使用微粒吸附仪(Micromeritics Gemini apparatus,ASAP 2020 M + C,Micromeritics,Co.,USA)进行测定,比表面积通过 BET 方法计算得到,总的孔隙率通过在相对压力为 0.97 条件下吸附氮的体积计算得到。

首先,EDTA 的接枝效率可以通过比较生物质修饰前后的总氮含量和羧基含量来评价,其结果如表 2.1 所示:化学修饰的生物质 CMTAB 的总氮含量(65.7 mg/g)大大高于未修饰生物质 OTAB($29.6\ mg\cdot g^{-1}$)。羧基的含量增加更为明显:OTAB 的羧基含量只有 0.297 $mmol\cdot g^{-1}$,而经过 EDTA 修饰之后,其羧基含量增加至 0.833 $mmol\cdot g^{-1}$,考虑到整个修饰过程除了 EDTA 本身,并没有其他氮源和羧基源,因此可以认为,EDTA 已经较为顺利地修饰到生物质的表面骨架结构中。

表 2.1　EDTA 修饰前后生物质中总氮和羧基功能团的含量比较

生物质	C_{COOH}(mmol·g^{-1})	总氮含量(mg·g^{-1})
OTAB	0.297	29.3
CMTAB	0.833	65.7

化学修饰前后生物质功能基团的变化情况通过红外光谱来表征。图 2.2 所示的是蒲草生物质修饰前后的红外光谱,根据谱图,下面将其中的红外吸收峰一一进行指认。在波数为 3440 cm^{-1} 左右的红外吸收峰是氨基或者羧基的伸缩振动峰,在 2926 cm^{-1} 和 2850 cm^{-1} 的峰分别被指认为是烷基 CH_2 的对称和非对称伸缩振动峰,而羧基、酰胺或者酯类中的 C=O 的伸缩振动峰是在 1732 cm^{-1} 和 1630 cm^{-1} 之间,在 1300 cm^{-1} 到 1000 cm^{-1} 之间的一系列吸收峰是 C—O 和 C—N 的伸缩振动或者羟基、氨基的弯曲振动峰,在 1000 cm^{-1} 以下的吸收主要是一些含氧无机离子(PO_4^{3-},SO_4^{2-} 等)的弯曲振动,以及 CH_2 的弯曲振动等[28]。比较修饰前后的红外谱图,发现其红外吸收峰所在的波数位置变化并不大,但是峰型有很大的改变,这主要是化学修饰过程导致的功能基团的化学状态发生了改变。此外化学修饰过后,在 2665 cm^{-1},1575 cm^{-1} 和 1185 cm^{-1} 处发现三个新的红外吸收峰,这可能是由于 EDTA 和生物质之间形成了新的化学键。

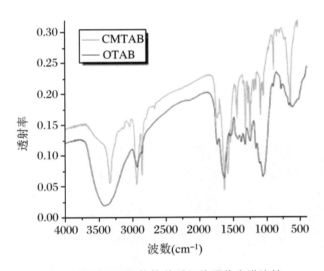

图 2.2　生物质 EDTA 修饰前后红外吸收光谱比较

表 2.2　生物质 EDTA 化学修饰前后功能基团的红外吸收分析

功能基团	波数(cm^{-1})	
	OTAB	CMTAB
OH 和/或 NH_2 伸缩振动	3415	3327
—CH_2 对称伸缩振动	2926	2929
—CH_2 不对称伸缩振动	2853	2850
酰胺和酯类中 C=O 伸缩振动	1732	1731
酰胺或者羧酸中 C=O 伸缩振动	1627	1626
—NH 弯曲变形	—	1575
CH_2 弯曲振动	1427	1436
C—N 伸缩振动	1318	1311
	1245	1243
	—	1185
C—O 伸缩振动	1157	1158
	1082	1088
	1057	1068
含氧无机离子(PO_4^{3-},SO_4^{2-})的伸缩振动	896	892
	780	782
	639	641

生物质被 EDTA 修饰前后的形貌通过扫描电子显微镜观察，如图 2.3 所示，化学修饰过程并未破坏生物质的基本骨架结构。生物质中的孔隙结构可以通过氮气吸/脱附曲线来表征，如图 2.4 所示，修饰前后生物质的孔隙结构都很少，基本可以认为是无孔隙结构，用 BET 方法测量的比表面积分别为 $0.778\ m^2 \cdot g^{-1}$ 和 $1.092\ m^2 \cdot g^{-1}$，进一步证实了修饰前后生物质都无明显的孔隙结构。

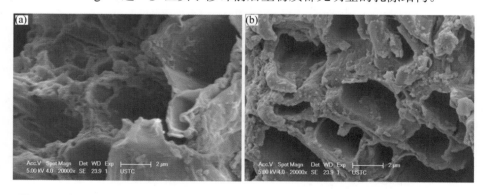

图 2.3　生物质化学修饰前后的表面形貌比较：(a) 修饰前；(b) 修饰后

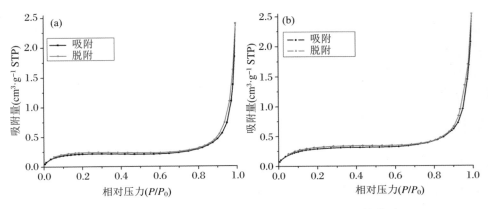

图 2.4 生物质修饰前后的氮气吸附-脱附曲线图:(a) 修饰前;(b) 修饰后

2.1.4
表面化学修饰生物质强化重金属去除性能解析

准确称取一定量(0.3 g)的吸附剂(OTAB 或 CMTAB)与 100 mL 的不同浓度(20~600 mg·L^{-1})的铅溶液混合在 250 mL 的锥形瓶中,然后置于水浴恒温振荡箱中,在常温下以 150 r·min^{-1} 的转速振荡 300 min,吸附过程的 pH 通过 1 mol·L^{-1} 的硝酸或者氢氧化钠溶液调节。当吸附过程结束时,将吸附混合物立刻抽滤,再将获得的固体(Pb-CMTAB 或者 Pb-OTAB)在 105 ℃ 下干燥去除水分,然后储存在干燥器中待用。滤液中剩余的铅元素浓度通过原子发射光谱进行分析(ICP-AES,Optima 7300 DV,Perkin Elmer Corporation,USA)。铅的去除效率(R)通过以下公式计算:

$$R = \frac{C_0 - C_e}{C_0} \times 100\% \tag{2.2}$$

其中,C_0 和 C_e 分别是初始和平衡时的铅浓度,而生物质的吸附能力(q_e)可以通过以下公式计算:

$$q_e = \frac{(C_0 - C_e) \times V}{W} \tag{2.3}$$

其中,W 和 V 分别为吸附剂的用量和铅溶液的体积,C_0 和 C_e 分别是初始和平衡时的铅浓度。

吸附完成之后,Pb-CMTAB 和 CMTAB 通过 X 射线光电子能谱(XPS)来分析表面功能基团的化学形态的变化情况。分析过程采用 XPS 能谱仪

(ESCALAB250，Thermo-VG Scientific，USA)，在恒定分析能量(constant analyzer energy)模式下进行测定。

图 2.5(a)所示是不同时间下 CMTAB 对重金属铅的去除情况，由图可知，CMTAB 对铅的吸附过程是一个快速平衡的过程，吸附平衡可以在 20 min 之内达到，此时的去除效率可以到达 85.9%。吸附动力学数据可以通过拟二级动力学模型进行模拟[29]：

$$\frac{t}{q_t} = \frac{1}{K_2 q_e^2} + \frac{1}{q_e} t \qquad (2.4)$$

其中，q_t 和 q_e 分别为时间 t(min)和平衡条件下 CMTAB 的铅吸附量(mg·g^{-1})，K_2 为动力学常数(g·mg^{-1}·min^{-1})。动力学参数通过对 t/q_t 和 t 作线性图，计算其截距和斜率而得。如图 2.5(b)所示，根据公式计算出动力学常数为 3.77×10^{-3} g·mg^{-1}·min^{-1}，其相关系数达到 0.998，证明拟二级动力学模型可以很好地模拟该吸附过程，理论计算得到的平衡吸附量为 47.62 mg·g^{-1}，和实验所测值相差不大，进一步证实了该模型的合理性，从而表明吸附过程的主要限速步骤为化学吸附[30]。

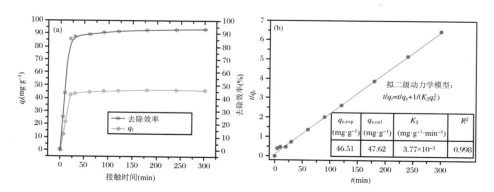

图 2.5 (a) 不同时间条件下 CMTAB 对铅的吸附能力和去除效率(铅的初始溶度为 100 mg·L^{-1}，初始 pH 为 5.0；生物质用量为 2.0 g·L^{-1}；温度为 25 ℃)；(b) 吸附过程的拟二级动力学模拟

图 2.6 所示是不同平衡溶度下，CMTAB 对铅的吸附能力，根据所得数据使用 Langmuir 和 Freundlich 等温线模型进行模拟[31-33]，求出 CMTAB 对铅的最大吸附能力。

$$\frac{C_e}{q_e} = \frac{C_e}{q_{max}} + \frac{1}{K_L q_{max}} \qquad (2.5)$$

$$\ln q_e = \ln K_F + \frac{1}{n} \ln C_e \qquad (2.6)$$

式中，K_L(L·mg^{-1})是与吸附自由能相关的 Langmuir 常数，q_{max}(mg·g^{-1})是

最大吸附量，C_e(mg·L^{-1})和 q_e(mg·g^{-1})分别是平衡铅浓度和吸附量；K_F(mg/g(L/mg)$^{-n}$)和 n 分别为与吸附能力相关的 Freundlich 常数和吸附强度。Langmuir 和 Freundlich 等温线模型的参数分别可以通过对 C_e/q_e 与 C_e 和 $\ln q_e$ 与 $\ln C_e$ 作线性曲线计算得到。根据图 2.6 所示，这两个等温线模型都能很好地模拟等温吸附过程，其相关系数都大于 0.95。相对而言，Freundlich 模型的相关系数为 0.99，大于 Langmuir 模型的 0.956，表明 CMTAB 对铅的吸附过程并不是一个严格的单层吸附过程，可能还涉及多层吸附。类似的现象在其他文献也有报道[32]。CMTAB 的铅最大吸附量为 263.9 mg·g^{-1}，远远大于未经修饰 OTAB 的吸附量(104.5 mg·g^{-1}，见图 2.7)，证明 EDTA 修饰过程对强化生物质去除重金属过程的有效性。

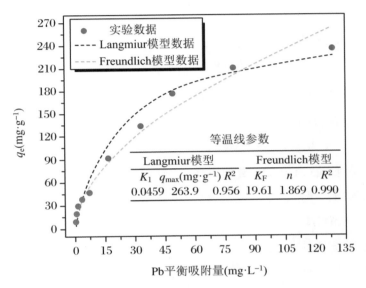

图 2.6　CMTAB 对铅的强化去除等温线

铅的初始溶度为 20～600 mg·L^{-1}，初始 pH 为 5.0；生物质用量为 2.0 g·L^{-1}；温度为 25 ℃；反应时间为 300 min

2.1.5
表面化学修饰对重金属强化去除机理研究

介质的 pH 是影响重金属去除机理的主要原因之一，它不仅可以影响生物质表面功能基团的化学状态，也能改变溶液中金属离子的存在形态[23-34]。以铅为例，在溶液 pH 为 2～8 时，铅离子有三种不同的存在形式：Pb^{2+}，$Pb(OH)^+$ 以

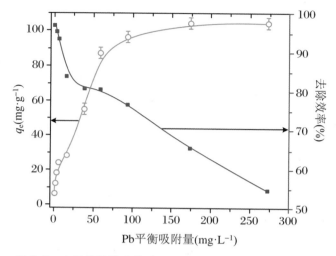

图 2.7 未经修饰的生物质 OTAB 对铅的去除等温线

铅的初始溶度为 20~600 mg·L^{-1},初始 pH 为 5.0;生物质用量为 2.0 g·L^{-1};温度为 25 ℃;反应时间为 300 min

及 Pb(OH)$_2$,其转化过程如以下方程所示:

$$Pb^{2+} + H_2O \longrightarrow Pb(OH)^+ + H^+, \quad \lg K_1 = -7.8 \quad (2.7)$$

$$Pb^{2+} + 2H_2O \longrightarrow Pb(OH)_2 + 2H^+, \quad \lg K_1 = -9.4 \quad (2.8)$$

不同铅的存在形式随 pH 的变化可以通过 MINEQL 软件计算[35],结果如图 2.8 所示,在 pH<4.0 时,铅主要是以 Pb^{2+} 的形式存在于溶液中,Pb^{2+} 水解形成 Pb(OH)$^+$ 和 Pb(OH)$_2$ 的过程开始于 pH=3.6,这个过程在 pH=5.0~7.0 这个窄区间内进展非常迅速,当 pH>6.0 时,Pb(OH)$_2$($K_{sp} = 1.2 \times 10^{-15}$)[36] 开始成为溶液中铅的主要存在形式了。

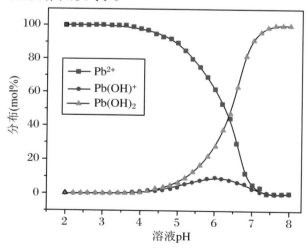

图 2.8 不同 pH 条件下,铅的三种存在形式在水溶液中的分布情况

为了研究不同 pH 对铅的去除机理的影响,我们通过改变 pH(2.0～8.0)进行了一系列实验,结果如图 2.9 所示。在初始 pH 在 2.0～6.0 之间时,吸附平衡时的 pH 显著升高,而当初始 pH 进一步升高至 7.0 或者 8.0 时,平衡时 pH 的变化不大,平衡吸附量从 pH=2.0 时的 19.7 mg·g^{-1} 迅速上升至 pH=5.0 时的 46.4 mg·g^{-1},并且在 pH 进一步升高的时候基本保持不变,这个趋势和文献报道的 EDTA 修饰生物质的基本一致[37],但是和未经 OTAB 修饰的生物质的变化趋势显著不同(图 2.10,OTAB 最大吸附量在 pH=6.0 时,并且随着 pH 进一步升高有明显的下降)。这种现象可以解释为:在低 pH 酸性条件下,铅离子和溶液中的大量 H$^+$ 离子将会竞争 CMTAB 表面的功能基团,从而导致了 CMTAB 对铅离子的固定能力下降。而且,在低 pH 条件下,CMTAB 表面的功能基团结合 H$^+$ 离子形成的质子化的结构和铅离子存在静电排斥作用,从而难以作为电子供体和铅离子形成配位作用,进一步降低了 CMTAB 对铅离子的固定作用。当溶液的 pH 上升时,一方面,H$^+$ 离子浓度降低,其和铅离子的竞争作用减弱,另一方面,CMTAB 表面功能基团的质子化作用降低,更容易和铅离子形成配位作用,在这些因素的共同作用下,铅离子的去除能力显著增加。

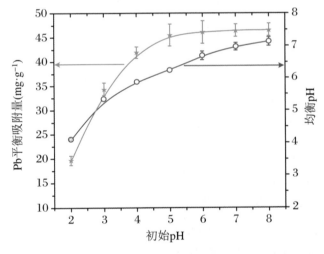

图 2.9　溶液初始 pH 对 CMTAB 去除铅能力的影响曲线

铅的初始溶度为 100 mg·L^{-1};生物质用量为 2.0 g·L^{-1};温度为 25 ℃;反应时间为 300 min

除了溶液的 pH,另一个对铅的去除机理影响很大的因素是表面功能基团。根据生物质的氮气吸/脱附等温线可知,生物质化学修饰前后都属于几乎无孔隙的结构(图 2.3),因此,主要由于孔隙结构导致的物理吸附过程基本可以忽略,铅的去除主要归因于生物质表面的功能基团的化学吸附。其中,对吸附起主要作用的化学基团主要包括氨基、羟基和羧基[9]。为了研究这三个主要功能基团

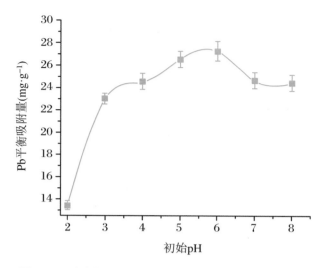

图2.10　溶液初始pH对OTAB对铅去除能力的影响曲线

铅的初始溶度为100 mg·L^{-1}，生物质用量为2.0 g·L^{-1}；温度为25 ℃；反应时间为300 min

在吸附过程中的具体贡献，我们进行了以下实验：通过化学反应对生物质的功能基团进行阻断，使其不能参与对铅的吸附过程，并和未阻断前的吸附量进行对比，从而得出功能基团对铅吸附的贡献。其中氨基和羟基的阻断反应如下方程所示：

$$2R\text{—}OH + HCHO \longrightarrow (R\text{—}O)_2CH_2 + H_2O \quad (2.9)$$

$$2R\text{—}NH_2 + HCHO \longrightarrow CH_2=CH_2 + H_2O \quad (2.10)$$

羧基的阻断反应如下方程所示：

$$2R\text{—}COOH + CH_3OH \longrightarrow RCOOCH_3 + H_2O \quad (2.11)$$

化学阻断前后生物质对铅的吸附能力如图2.11所示，从中可以看出，无论哪个基团被阻断，生物质对铅的吸附能力都下降明显。其中羟基和氨基阻断后，生物质对铅的吸附能力从26.94 mg·g^{-1}下降到17.03 mg·g^{-1}，下降比例为36.70%，而羧基阻断后，吸附能力下降更为显著，从26.94 mg·g^{-1}下降到12.42 mg·g^{-1}，下降比例超过50%，证明羧基对铅的吸附能力贡献要显著大于氨基和羟基，这和文献报道的结果也是一致的[20,38]。

一般认为，羧基、氨基、羟基等功能基团对重金属离子的吸附作用主要是通过离子交换和配位作用来实现的：

$$R^{n-}E^{n+} + M^{n+} \leftrightarrow R^{n-}M^{n+} + E^{n+} \quad \text{（离子交换作用）} \quad (2.12)$$

$$M^{n+} + mBL \leftrightarrow M(BL)_m^{n+} \quad \text{（配位作用）} \quad (2.13)$$

其中，$R^{n-}E^{n+}$表示生物质表面可用于离子交换的活性位点，M^{n+}代表金属离子，

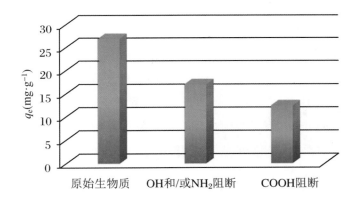

图 2.11　功能基团阻断前后生物质对铅的吸附能力的比较

铅的初始溶度为 100 mg·L^{-1}，生物质用量为 3.0 g·L^{-1}；温度为 25 ℃；反应时间为 300 min，pH=6.0

E^{n+} 是生物质表面用于交换的离子，通常为氢离子或者是钾、钠离子等。BL 表示生物质表面可用于配位的基团（一般是含氨基、羧基或者羟基等）。除了离子交换和配位作用，氢键作用也被认为在一定 pH 条件下可以影响吸附的化学作用：根据之前的分析，在 pH>4.0 时，溶液中开始存在 Pb(OH)$^+$ 和 Pb(OH)$_2$，其中的氢氧根可以和生物质表面的氨基、羟基和羧基形成氢键相互作用。因此，一个 pH 依赖的，耦合了氢键、离子交换和配位作用的机理，可以更加合理地解释表面化学修饰生物质对铅的强化去除作用。吸附机理示意如图 2.12 所示。

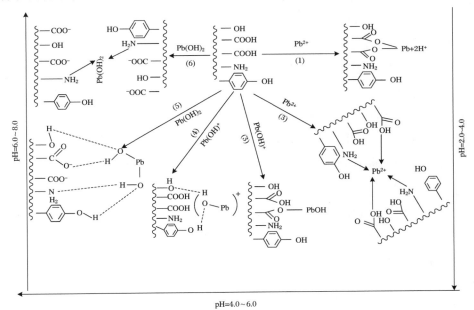

图 2.12　表面化学修饰生物质强化铅去除的机理示意图

在 pH=2.0~4.0 时，铅离子主要通过和生物质表面羧基的氢离子进行离子交换而被吸附，同时，由于氨基或者羟基中的氮氧原子存在空的电子轨道，也可以提供给铅离子，形成配位作用，但是由于低的 pH 容易使得这两个功能基团质子化，从而和铅离子之间产生静电排斥，因此，配位作用在低的 pH 条件下将会显著削弱。当 pH 升高至 4.0~6.0 之间时，溶液中铅的存在形式包括 Pb^{2+}、$Pb(OH)^+$ 和 $Pb(OH)_2$，因此吸附机理也变得复杂了很多，除了有离子交换、配位作用之外，氢键作用也开始形成了。当 pH 进一步升高至 6.0~8.0 时，溶液中的铅离子消失，$Pb(OH)_2$ 开始成为主要存在形式，吸附机理主要变成了氢键和配位作用，离子交换作用基本消失。

我们通过吸附过程前后溶液 pH 的变化，估算了离子交换、氢键和配位作用对吸附过程的贡献，其计算公式如下：

$$q_{离子交换} = \frac{([H^+]_{Pb} - [H^+]_0) \times V \times M_{Pb} \times 0.5}{W} \quad (2.14)$$

$$\eta_{离子交换}(\%) = \frac{q_{离子交换}}{q_e} \times 100\% \quad (2.15)$$

$$[H^+] = 10^{-pH_e} - 10^{-pH_i} \quad (2.16)$$

其中，$q_{离子交换}$ 是指通过离子交换作用吸附的铅离子量，$[H^+]_{Pb}$ 和 $[H^+]_0$ 分别为吸附过程和控制实验中氢离子的释放量，V 是溶液的体积，M_{Pb} 是铅的相对分子质量，0.5 是氢离子和铅离子的电荷比，W 是吸附过程所用生物质的量，$\eta_{离子交换}$ 是离子交换作用在铅的总吸附量中所占比例，pH_e 和 pH_i 分别为平衡和初始的 pH 值。配位作用和氢键作用的贡献比较复杂，难以分开，可以合在一起计算，其值等于 $100 - \eta_{离子交换}$。

图 2.13 所示是不同化学作用对铅的去除作用的贡献随 pH 的变化，从图中可知，pH 在 2.0~3.0 之间时，离子交换作用是对铅吸附做最主要的贡献，占到 70% 以上。当 pH 上升时，离子交换作用持续减弱，到 pH>7.0 时趋近于 0，而配位和氢键作用则不断增强，和之前的机理分析是一致的。

最后，我们通过 XPS 分析生物质在铅的吸附前后表面功能基团的化学状态变化，进一步阐述化学修饰生物质对铅的强化去除机理。图 2.14 所示的是 CMTAB 吸附铅前后的 XPS 图谱，其中图 2.14(a) 是 XPS 全谱扫描图，结合能在 1072 eV 和 497 eV 的峰被指认为 Na 1s 和 Auger 光电子激发峰。但是这两个峰在铅吸附后消失了，这主要是在吸附过程中铅离子和钠离子发生了离子交换，铅离子取代了钠离子在生物质上的位置，相似的现象在其他文献中也有报道[39-40]。CMTAB 的 C 1s XPS 光谱可以分成 4 个峰（图 2.14(b)），这些峰可以分别指认为：C—C（石墨态、无定形态，284.6 eV）；C—O（醇、酚羟基以及醚类，

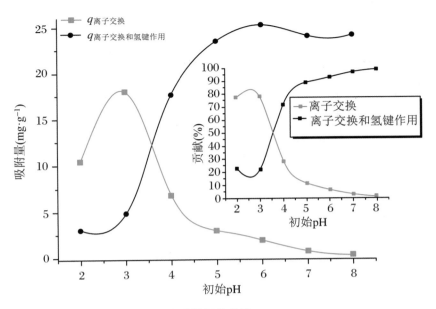

图 2.13 不同化学作用对铅去除的贡献估算

286.2 eV);C=O(羰基和醌类,287.85 eV);O=C—O(羧酸或者酯类,289.45 eV)[41-43]。其中,C—C,C—O,C=O 的峰在铅吸附后基本没有变化,而 O=C—O 的峰消失了,这可能是由于吸附过程中形成了羧基和铅的配位化合物[44],因此可以认为,羧基是生物质吸附铅的主要贡献基团。图 2.14(c)所示为 CMTAB 吸附前后 O 1s XPS 光谱,CMTAB 中 O 1s 的结合能为 532.75 eV,而 Pb-CMTAB 的 O 1s 的结合能下降为 532.52 eV(结合能的下降主要是由于氧原子的多余电子和铅原子进行配位,从而降低了氧原子周围的电子云密度,使得其结合能下降,相同的趋势在 N 1s XPS 光谱也观察到了(图 2.14(d)),表明氮原

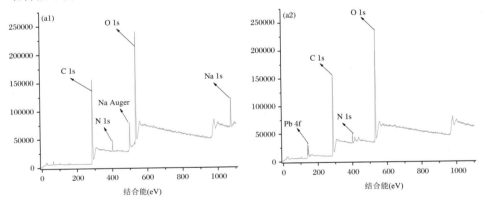

图 2.14 CMTAB 和 Pb-CMTAB 的 XPS 光谱:(a) XPS 全谱扫描(a_1:吸附前,a_2:吸附后);(b) C 1s 光谱(b_1:吸附前,b_2:吸附后);(c) O 1s 光谱;(d) N 1s 光谱;(e) Pb 4f 光谱

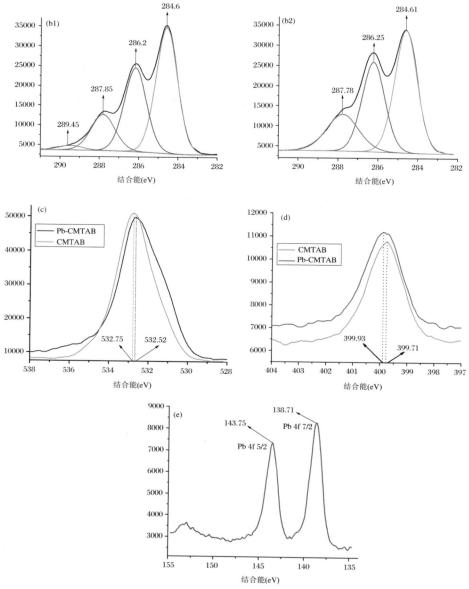

续图 2.14　CMTAB 和 Pb-CMTAB 的 XPS 光谱：(a) XPS 全谱扫描(a_1：吸附前，a_2：吸附后)；(b) C 1s 光谱(b_1：吸附前，b_2：吸附后)；(c) O 1s 光谱；(d) N 1s 光谱；(e) Pb 4f 光谱

子也可以和铅原子形成类似的配位作用。图 2.14(e)所示的是 Pb-CMTAB 中铅原子的 4f XPS 光谱，其中结合能在 138.71 eV 的峰表明铅原子是和羧基配位结合的[45]。

本节研究了通过使用二氯亚砜活化的 EDTA 作为修饰试剂，对蒲草生物质进行表面化学修饰，使其含有丰富的羧基功能基团，然后将其用于去除废水中的重金属铅，获得以下结论：

（1）二氯亚砜活化的 EDTA 是一个对蒲草进行表面化学修饰的有效试剂，经过修饰之后，生物质表面的羧基含量显著增加。

（2）经过修饰的生物质对水溶液中的铅离子具有显著强化的去除能力，其最大吸附能力达到 263.9 mg·g^{-1}，远远大于未经修饰的生物质的 104.5 mg·g^{-1}。

（3）铅的吸附过程是一个快速的过程，吸附平衡可以在 20 min 内达到，吸附动力学符合拟二级动力学模型表明化学吸附是主要的限速步骤。

（4）吸附过程是一个高度依赖 pH 的过程，pH=5.0 时吸附效果最佳。根据 XPS 和其他相关分析，吸附机理主要涉及离子交换、配位和氢键作用，其中离子交换作用主要是在低 pH 条件下起作用，而配位和氢键作用是在高 pH 条件下起作用。

2.2 铅污染生物质热解铅元素迁移转化过程及机理

2.2.1 概述

生物质吸附作为一种低成本、易操作的方法被广泛用于废水中重金属的去除[46-54]。然而，该过程中使用的生物质在吸附完成后通常是被重金属严重污染的，如果处置不当容易造成二次污染。处置这些重金属污染生物质的常规方法是通过强酸溶液（HCl、H_2SO_4 或者 HNO_3）或者 EDTA 溶液对其进行浸取，使得生物质中的重金属重新脱附并且回收[26,55-57]。这种方法尽管可以回收大于 95% 的重金属并且吸附剂可再生，但是也存在一些缺点：首先，浸取过程需要消耗大量的浸出溶液（通常 1 g 重金属污染的生物质需要消耗 20 mL 的浸出溶液），从而抵消了生物质吸附过程低成本的经济优势；其次，使用过的浸出溶液的排放也容易造成二次污染或者增加后续的处理成本。因此，迫切需要寻找一种低成本、易操作的绿色方法处置这些重金属污染生物质。快速热解作为一种广泛用于从废弃生物质中回收能源和资源的方法，已经被证明可以用来处置那些含有用于植物修复后含重金属的生物质[58-61]。由于其易操作性和高的能源回收效率，快速热解应该可以作为一种有前景的方法用于处理重金属污染的生物质[17,62-66]。在本节研究中，我们采用蒲草生物质去除废水中的重金属铅，将获得的铅污染生物质再进行快速热解，回收生物油资源和重金属铅，研究重金属铅在热解过程中的迁移转化过程及机理，并将此方法和常规的溶液浸取方法进行成本对比。

2.2.2
铅污染生物质的快速热解过程及其产物分析

蒲草生物质采集自南淝河。采集回来以后，先用水洗净，再切成小段，然后在 120 ℃下干燥除水，之后将干燥过的生物质粉碎并过筛，收集粒径小于 100 目的颗粒，储存起来备用。实验中所用的化学试剂都为分析纯。快速热解过程是在一个自由落体式的石英管式反应器（图 2.15）中实现的，石英管的长度为 600 mm，内径为 23 mm。石英管置于一个大功率管式电炉中，其温度通过一个 K 形的热电偶来控制。热解开始前，将 4.0 g 的生物质放置于进料管中，当温度达到设定值时，开启载气流（$0.4 L \cdot min^{-1}$ 的氩气），保持 20 min 以排尽热解系统中的空气，随后，将生物质样品通过一个活塞送入反应器中进行热解。热解产生的挥发性物质通过氩气载气吹出，其中的可冷凝的物质通过冷的乙醇-乙二醇混合物（温度为 -20 ℃）冷凝，获得生物油，不可冷凝的气体则通过气袋收集。热解过程结束后，石英管反应器立刻从电炉加热区移出，在氩气氛围下冷却至室温。对获得的生物油和生物炭进行称重并计算其产率。

对热解产生的生物油和生物炭中的碳、氢、氧、氮元素含量使用元素分析仪（VARIO EL Ⅲ，Elementar Inc.，Germany）进行测定。铅的测定过程是通过先对生物油和生物炭样品进行化学消解（使用浓硫酸-过氧化氢体系在 350 ℃ 消解至无色透明液体），再通过原子发射光谱（ICP-AES，Optima 7300 DV，Perkin Elmer Co.，USA）进行测定。生物炭中铅的化学形态通过 X 射线衍射（XRD，MXPAHF，Rigaku，Japan）进行分析。铅污染生物质的浸出实验采用 EDTA（$0.1 mol \cdot L^{-1}$）和盐酸（$0.1 mol \cdot L^{-1}$）溶液作为浸出液。在每个浸出实验中，将 0.5 g 的铅污染生物质和 50 mL 的浸出溶液加入锥形瓶中，置于恒温振荡箱

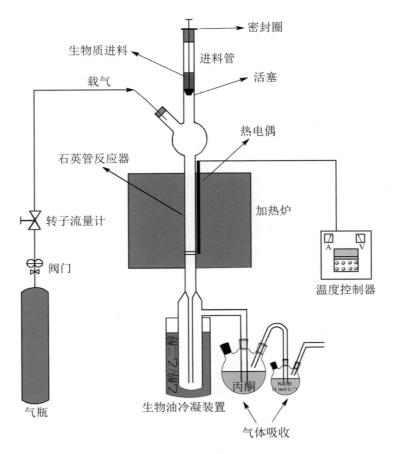

图 2.15　自由落体式石英管反应器示意图

中,在 25 ℃下以 150 r·min^{-1}的转速振动搅拌 24 h,然后将混合液分离,并用 ICP-AES 测定液体中的铅含量,计算铅的回收效率。

首先,我们研究了温度对热解产物的产率和组成的影响。一系列的热解实验在不同温度(400~600 ℃)下开展,测定了热解过程中的生物油、热解气和热解生物炭的产量,并计算其各自的产率。随着热解温度从 400 ℃上升到 600 ℃,生物炭的产率迅速从 46.5%下降到 33.0%,而热解气的产率则从 15%左右升高到接近 30%,这个结果和有关文献报道基本一致[67]。生物炭产率迅速下降的主要原因是热解温度的升高导致的生物质持续分解。而热解气产率的升高则是由于热解生成的挥发组分在温度升高时发生了二次裂解,生成了小分子的其他物质。对于生物油,其产率在 500 ℃时达到最大值,随后下降,这主要是因为在低温的时候,生物质中的挥发组分不能得到充分的分解,而温度的升高有利于其分解,从而提高生物油的产率。当温度达到一定值时,进一步升高温度又将导致挥发组分进一步裂解成小分子气体,从而使得生物油产率降低。因此,生物油的产率存在一个峰值(500 ℃)。不同温度下获得的生物油和生物炭的元素组成如表 2.3 所示，根据其元素组成，使用以下公式，可以计算出热值(HHV)：

$$HHV(MJ \cdot kg^{-1}) = (3.55C^2 - 232C - 2230H + 51.2CH + 131N + 20600)/1000 \tag{2.17}$$

式中，C 为 C 的含量，H 为 H 的含量，N 为 N 的含量，O 为 O 的含量。

表 2.3 不同温度下热解铅污染生物质(Pb-TAB)获得的生物油和生物炭的元素组成和热值

	温度（℃）	元素组成（wt%）				HHV（MJ·kg^{-1}）
		C	H	N	O[①]	
生物炭[②]	400	61.27	3.30	3.79	31.64	23.20
	500	64.93	3.41	3.37	28.59	24.68
	600	67.22	3.53	3.00	26.25	25.71
生物油	400	28.17	9.66	1.48	60.69	9.47
	500	26.39	9.72	1.60	62.29	8.62
	600	24.89	9.79	1.79	63.53	7.90
Pb-TAB	—	42.23	6.01	2.67	49.09	17.08

注：① 通过差减法计算。
② 不含灰分。

由此可知，随着热解温度的升高，生物炭的热值不断升高，而生物油的热值呈现降低的趋势。

2.2.3

铅元素在热解过程中的迁移转化过程及机制解析

接下来，我们研究了重金属铅在热解过程中的迁移转化过程。图 2.16 所示不同热解温度下铅在热解产物中的分布情况，由图可知，随着温度的升高，铅在生物油和生物炭中的含量都呈现不断上升的趋势。在生物油中，铅的含量从 400 ℃时的 2.9 mg·kg^{-1}上升到 600 ℃时的 13.8 mg·kg^{-1}。尽管铅含量有成倍的上升，然而其实际含量仍然不高于国家规定的无铅汽油中铅的含量，因此，当生物油经过精制用于燃料时，其导致的铅的二次污染基本可以忽略。对于生物炭，其中铅的含量从 400 ℃时的 5.39%上升到 600 ℃时的 7.53%，这个含量已经接近于常规铅矿石中铅的含量[68]，可以比较方便地提取其中的铅元素。热解过后，生物炭中铅的存在形态通过 XRD 进行表征，其结果如图 2.17 所示。

其中在衍射角度 2θ 为 31.5°,36.2°,40.7°和 62.4°的峰是金属铅的衍射峰,而衍射角度 2θ 为 28.6°,50.4°和 54.1°的峰被认为是氧化铅(PbO)的峰。根据 XRD 结果,可以知道,生物质负载的铅离子在热解过程中发生了一系列的分解和还原过程,其可能经历的反应过程推测如下:

$$Pb(NO_3)_2 + 2H_2O \longrightarrow Pb(OH)_2 + 2HNO_3 \uparrow \quad (干燥过程) \quad (2.18)$$

$$Pb(OH)_2 \longrightarrow PbO + H_2O \uparrow \quad (热解过程) \quad (2.19)$$

$$PbO + C(CO, H_2) \longrightarrow Pb + CO(H_2O, CO_2) \uparrow \quad (热解过程) \quad (2.20)$$

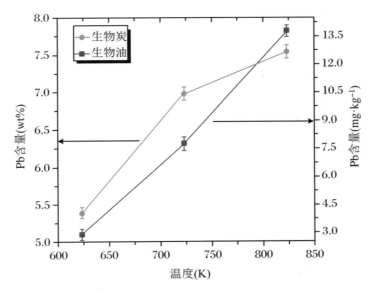

图 2.16 不同热解温度下热解产物中铅含量变化

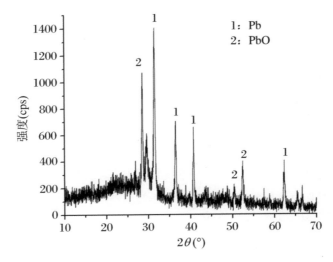

图 2.17 生物炭中铅元素的 XRD 图谱

首先，生物质吸附的硝酸铅在干燥过程中水解生成氢氧化铅；随后，在热解过程中，氢氧化铅发生分解产生氧化铅；最后，一部分氧化铅被生物质热解过程中生成的还原组分还原成金属铅。

2.2.4
铅污染生物质快速热解方法的可持续性影响解析

为了研究快速热解方法对于处理重金属污染生物质的技术经济适用性，我们通过比较快速热解方法和常规的溶液浸取方法对于处理同样的铅污染生物质的成本和铅回收效率来对其进行技术经济评价。首先，比较这两种方法对铅的回收效率。对于快速热解过程，在生物炭中富集的铅被认为是可回收的铅，其回收效率 $R_{pyrolysis}$ 可以通过以下公式计算：

$$R_{pyrolysis}(\%) = \frac{\text{生物油中铅的含量}\left(\frac{mg}{g}\right) \times \text{生物油的质量}(g)}{\text{铅污染生物质中铅含量}\left(\frac{mg}{g}\right) \times \text{生物质的质量}(g)} \times 100\% \tag{2.21}$$

而常规浸出方法铅的回收效率 $R_{leaching}$ 通过以下公式计算：

$$R_{leaching}(\%) = \frac{\text{总的可浸出铅的质量}(mg)}{\text{铅污染生物质中的铅含量}\left(\frac{mg}{g}\right) \times \text{生物质的质量}(g)} \times 100\% \tag{2.22}$$

这两种方法对铅的回收效率如图2.18所示，EDTA作为一种最常用的配位试剂，可以回收铅污染生物质中95.5%的铅，而盐酸溶液的回收效率也基本相同，达到94.9%。而在热解过程中，尽管随着温度的升高，铅的回收效率会稍微降低，但是其整体回收率在600 ℃时仍然能达到98%以上，表明快速热解过程确实能够有效地回收受污染生物质中的重金属，并且温度对铅的回收效率影响不大，因此我们可以放心选择合适的温度使得热解过程中生物油的产率和性质最佳，而不用担心由此导致的铅的回收率降低。

快速热解方法处理铅污染生物质的经济评价首先设定以下基本条件：吸附过程中生物质用量为每吨废水3.0 kg，铅含量为100 mg·L^{-1}，所有的经济参数是根据中国市场2010时基本经济情况来估算的。基本的流程如图2.19所示，生物质预处理包括收割、干燥和运输过程。由于蒲草生物质是一种广泛存在于

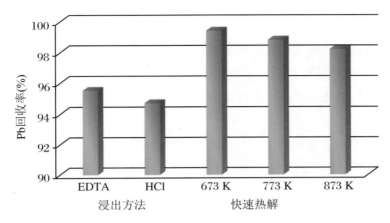

图 2.18　快速热解和浸出方法对铅污染生物质中铅的回收效率比较

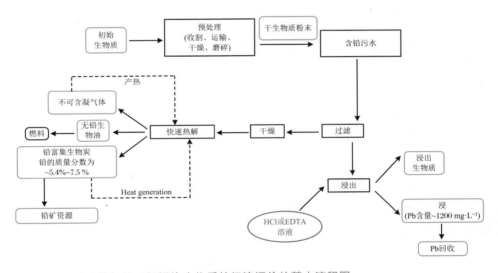

图 2.19　快速热解处理铅污染生物质的经济评价的基本流程图

结构湿地中的水生植物，其本身价值可以忽略，收割和运输成本（运输距离在 20 km 以内）为每吨生物质 200 元，干燥过程是从含水率 25%（自然晒干含水率）干燥至 7%（热解进料含水率），其成本估计为每吨生物质 35 元。生物质的粉碎过程成本按照其消耗能量来估算：每吨生物质粉碎至粒径为 0.8 mm 时需消耗电能 27.6 kWh，成本为 16 元，吸附过程和随后的过滤成本估算为每吨废水 0.2 元，经过换算，整个吸附过程的成本为每吨废水 0.92 元。热解过程中生物油的最大产率为 45.7%，表明每吨生物质可以产出 457 kg 的生物油，其热值为 3939 MJ（8.62 MJ·kg^{-1}），等于 93.8 kg 柴油的热值。这些生物油的价值估算为 620 元。由于热解过程是一个基本可以实现自己供能的过程（热解产生的气体和生物炭可以作为能量提供热解过程），因此过程中的能耗可以忽略。但是由于热解设备投资较大，其折旧成本不可忽略，估算为每吨生物质75元，生物质吸附后的干燥过程是从过滤后的生物质含水率78%干燥至热解进料的7%，其成本为每吨生物质350元。考虑到热解过程产生的生物油可以获得620元价值，因此整个热解过程反而是盈利的，其盈利经过换算为每吨废水0.55元。减去吸附中每吨废水0.92元的成本，整个吸附热解过程的成本为每吨废水0.37元。

为了比较，我们也估算了常规的浸出方法的经济成本。在浸出过程中，设备投资相对来说会小很多，其设备的折旧费基本可以忽略，我们采用盐酸和EDTA作为浸出溶液来计算。浸出溶液和污染生物质的比例为20∶1。盐酸（体积比36.5%）和EDTA的工业成品价格分别为每吨370元和15000元。每吨废水需要消耗3 kg的生物质吸附剂，随后的解吸过程需要60 kg的0.1 mol·L^{-1}的浸出溶液。经过换算，每吨废水消耗的浸出试剂成本为0.25元（盐酸）和26.4元（EDTA），因此整个吸附浸出过程的成本为1.17元（盐酸）和27.3元（EDTA）。这个值显然比快速热解大很多，因此，快速热解作为一个处理重金属污染生物质的方法，具有很大的经济优势。

在本节研究中，我们采取快速热解的方式对生物吸附过程中产生的重金属污染的生物质进行处理，回收其中的生物油和重金属铅。根据实验结果，可以得出以下结论：

（1）温度是影响热解产物的产率和性质的主要因素，热解过程中生物油的最大产率为45.7%，其温度为500 ℃。生物油和生物炭中铅的含量随着温度的升高而不断升高。

（2）热解过程中铅的回收率受温度的影响不是很明显，整个热解过程，温度从400 ℃升高至600 ℃时，铅的回收率在98.8%以上，主要原因是铅的化合物在热解过程中转化为挥发性很弱的氧化铅和金属铅。

（3）技术经济比较表明，热解过程相比于常规的浸出过程，既可以更加有效地回收重金属铅，也可以显著降低成本。

2.3 湿地植物生物质热解氮磷营养元素迁移转化过程及其机制解析

2.3.1 概述

水体富营养化是目前较受关注的环境问题之一，其最主要的成因是大量氮磷营养元素被排放至水体中[69]，因此，去除水体中的氮磷是控制水体富营养化的主要方法之一。目前应用于去除水体氮磷营养元素的方法很多[70-73]，其中通过种植人工湿地植物，利用其超高的氮磷富集能力，在其生长周期内大量吸收氮磷从而达到去除水体中氮磷的目的，被认为是一种有效的、低成本的去除手段，并且已经得到广泛的应用[74-83]。在一些结构湿地中，20%～30%的氮磷都是由植物的吸收而得到去除的[84]。湿地植物摄取氮磷主要用于合成其生长所必需的蛋白质以促进种子的萌发和根的发育等[85]。在生长季节，植物摄取大量的氮磷用于构建其主体，产生了大量的生物质，然而在其衰亡季节，这些生物质将会枯萎和腐烂，其氮磷营养元素将会重新释放到水体之中[86-87]。在其衰亡期收割这些已经停止生长的生物质是防止其产生二次污染的主要方法。之前的研究表明，在适当的时机收割这些生物质不仅能够有效地防止氮磷的再释放，还可以最大限度地回收这些营养元素[88-89]。收割回来的生物质可以用于发酵产甲烷和氢气[90]，也可以经过加工用作动物饲料[91-92]。然而，时间、劳动力和运输成本高及储存特性差限制了其广泛的应用。

快速热解是一个可以用于大规模处置这些收割下来的生物质的方法。一方面，它可以将生物质转化为液体生物油，用作化石燃料的替代品或者提取高附加值化学品[93]。另一方面，相对于生物质本身，生物油和生物炭的能量密度和储存性能显著提高，方便储存和运输。蒲草、水花生（*Alligator weed*）和水芹（*Oenanthe javanica*）是三种常见的结构湿地植物，它们都是一年生的植物，生长周期短、生长速率快、氮磷富集能力高，因此被广泛用于去除水体中的氮磷营养元素[94]。虽然生物质热解技术得到了广泛研究,但研究人员主要关注热解油过程及产品品质,尚无有关热解过程中氮磷营养元素的研究。在本节的研究中,我们选择了这三种具有代表性的湿地植物,研究了其热解过程中氮磷营养元素的分布情况和迁移转化规律,探讨了热解过程中温度对氮磷分布与转化的影响,通过特定的浸出实验及氮磷在生物炭中的形态分析,探索了回收氮磷的可能性。

2.3.2
湿地植物生物质快速热解及其产物分析

蒲草、水花生和水芹采集自一条严重富营养化的污染河流——南淝河。采集回来以后,先用水洗净,再切成小段,然后在120 ℃下干燥除水,之后将干燥过的生物质粉碎并过筛,收集粒径小于100目的颗粒,储存起来备用。

生物质原料的主要元素组成(碳、氢、氧和氮)通过元素分析仪(VARIO EL Ⅲ,Elementar Inc.,Germany)测定,其他的无机成分经过化学消解后通过原子发射光谱(ICP-AES,Optima 7300 DV,Perkin Elmer Co.,USA),其含量以氧化物的形式给出。生物油和生物炭的主要元素组成也是通过元素分析仪来测定的。磷的测定过程是先氧化消解,然后进行显色,最后通过紫外可见分光光度计测定,其主要步骤如下:精确称量的 0.1 g 的生物油或生物炭样品,通过浓硫酸和过氧化氢在 350 ℃消解至无色透明,然后定容至 100 mL 的容量瓶中,随后取出一定量的消解液,依次加入钼酸铵、酒石酸锑钾和 L-抗坏血酸显色至蓝色,然后在 700 nm 的条件下测试其吸光度,经过计算得到样品中总磷的含量。

生物油中的组分的化学结构通过气相色谱-质谱联用仪(GC-MS)来测定。测定之前,生物油先通过二氯甲烷进行萃取,获得的有机相通过 GC-MS 分析其主要成分的结构。GC-MS 的仪器由气相色谱(Agilent-7890A)和质谱(Agilent-975C)组成,使用 HP-5MS 非极性毛细管柱(30 m-0.25 mm-0.25 μm)。使用 1.0 mL·min^{-1} 的高纯氮气作为载气,分流比为 1:20。测试过程采取程序升温的模式,毛细管柱温首先以 4 ℃·min^{-1} 从 40 ℃升高至 180 ℃,然后以 10 ℃·min^{-1} 升高至 250 ℃,在这个温度下保持 5 min,然后开始降温。检测器的温度保持 250 ℃不变。

生物炭中的氮磷的化学状态通过不同的浸出液的浸出实验来表征。铵态氮(NH_4^+-N)和硝态氮(NO_3^--N)通过氯化钾(KCl)和硫酸钾(K_2SO_4)的浸出来进行区分。而磷的化学状态可以通过碳酸氢钠($NaHCO_3$)和氟化铵/盐酸(NH_4F/HCl)

溶液的浸出来进行区分。其中碳酸氢钠浸出的磷是植物可以快速利用的磷,而氟化铵/盐酸浸出的磷是与铁、铝、钙结合的磷。这两种磷都被认为是植物可利用的磷。浸出过程的基本步骤如图2.20所示。

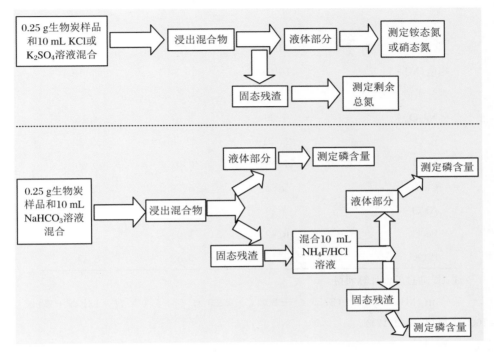

图2.20 生物炭中氮磷的浸出过程示意图

首先,将热解进料湿地植物的基本组成和特性进行详细的分析,结果如表2.4所示,本研究所用湿地植物的灰分含量在10.93%～17.25%之间,比常见的木质生物质的灰分(通常小于3%)大了很多[95],其组成主要是钾、钙、镁、钠、铝、磷等无机组分。此外,由于湿地植物优秀的氮磷富集能力,其生物质中氮磷含量也比常规木质纤维类生物质大为增加。

表2.4 热解进料湿地植物的基本组成和特性

	水花生	水芹	蒲草
近似分析(wt%)			
水分	2.81	6.29	4.42
挥发性物质	57.74	65.62	72.02
固定碳	25.01	18.14	17.05
灰分	17.25	16.24	10.93
元素分析(wt%)			
C	38.95	40.14	42.30

续表

	水花生	水芹	蒲草
H	5.24	5.26	5.69
N	2.42	3.42	2.93
P	0.26	0.47	0.58
O[①]	53.14	50.71	48.50
热值(MJ·kg^{-1})[②]	16.03	16.54	17.16
灰分组成(wt%)[③]			
Na$_2$O	4.35	3.67	1.16
K$_2$O	24.31	41.48	35.32
MgO	2.58	3.08	3.90
CaO	5.88	5.39	12.85
Al$_2$O$_3$	6.29	2.13	1.06
Fe$_2$O$_3$	0.79	1.01	0.43
P$_2$O$_5$	4.18	5.52	8.11

注：① 通过差值计算而得。

② 热值(MJ·kg^{-1}) = $(3.55\,C^2 - 232\,C - 2230\,H + 51.2\,C \times H + 131\,N + 20600)/1000$[96]（见公式(2.17)）。

③ 通过元素含量计算而得。

在热解过程中，温度是影响热解产物（生物油、生物炭和热解气）产率和化学组成的最主要因素[97]。图 2.21 所示是不同温度条件下热解产物的产率分布。从图中可以看出，对于三种湿地植物，随着温度从 723 K 升高到 873 K 时，其热解气体的产率从 14.5% 迅速增加到 29.8%，而生物炭的产率呈现不断下降的趋势；相对热解气和生物炭而言，生物油的产率变化趋势更为复杂，其产率呈现出先增加，后降低的趋势，其最高产率出现在 723 K 或者 773 K（蒲草生物油产率最高为 43.6%，723 K；水花生生物油产率最高为 42.3%，723 K；水芹生物油产率最高为 40.6%，773 K）。这主要是由于随着温度的升高，木质素、纤维素等不断降解成挥发性物质，冷凝成生物油，而随着热解温度的进一步升高，挥发性物质继续裂解产生小分子气体（氢气、甲烷、乙烯等），导致生物油的产率的降低和热解气体的增加。这个趋势和文献报道的结果是一致的[67,98-99]。

为了进一步研究湿地植物热解产物的组成和性质，我们分析了在生物油最高产率条件下获得的生物油和生物炭。表 2.5 所示是生物油和生物炭的元素组成和热值，从表中可知，湿地植物生物油的热值都比常见的木质生物质所产生物油的热值低[93]，这主要是由湿地植物的氧含量较高、碳氧比较低导致的[100]。此外湿地植物生物油和生物炭的氮磷含量显著高于常规生物油(通常小于 0.2%)[93],这主要是由湿地植物进料的高氮磷含量导致的。

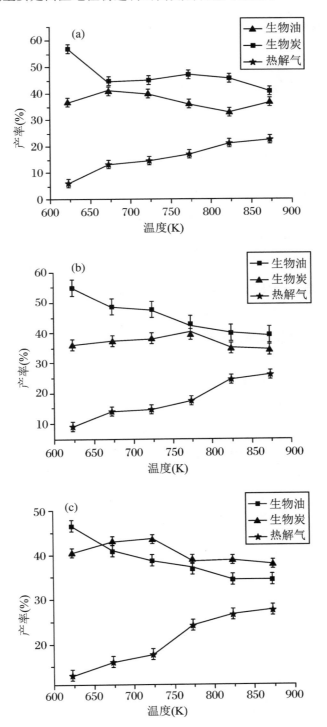

图 2.21 不同温度下,湿地植物热解产物的产率分布:
(a) 水花生;(b) 水芹;(c) 蒲草

表 2.5 湿地植物生物质热解产物的元素组成和热值

元素 (wt %)	水花生		水芹		蒲草	
	生物炭	生物油	生物炭	生物油	生物炭	生物油
C	54.55	23.49	56.82	25.08	61.97	27.57
H	2.06	9.09	2.14	9.67	2.98	9.13
N	3.28	1.92	4.04	2.48	3.99	2.77
P	0.48	0.07	0.93	0.16	1.07	0.15
O*	39.63	64.78	36.07	62.61	29.99	60.38
热值(MJ·kg^{-1})	20.09	8.02	20.87	8.19	23.19	9.79

注：* 通过差值法计算。

尽管所得生物油的热值较低，氧含量较高，不适合直接作为燃料使用，然而，由于湿地植物生物油的独特组成和特性，其作为原料提取高附加值化学品是一个不错的选择。表 2.6 所示是 GC-MS 分析的生物油主要化学成分的组成和相对含量。结果显示，湿地植物生物质是一个非常复杂的混合物体系，包含大量的含氧有机化合物，例如酚类、羧酸和醛酮类。生物油中化合物根据其结构可以大致分为三类：① 苯酚及其衍生物，其总含量分别为 16.65%（水花生生物油）、19.06%（水芹生物油）和 19.46%（蒲草生物油）；② 其他含氧有机物，这是生物油中含量最高的组分，包括羧酸、醛酮类等，其总含量分别为 55.98%（水花生生物油）、44.36%（水芹生物油）和 40.92%（蒲草生物油）；③ 含氮有机物，主要是含氮杂环化合物和胺类物质，其含量分别为 14.83%（水花生生物油）、23.00%（水芹生物油）和 17.92%（蒲草生物油）。其中的杂环化合物，例如吡啶、吡咯及其衍生物主要来源于生物质中生物碱和叶绿素等的裂解，而胺类物质主要来源于氨基酸的裂解。

表 2.6 湿地植物生物油中主要组分的结构和相对含量

		水花生		水芹		蒲草	
		主要组分	相对含量（%）	主要组分	相对含量（%）	主要组分	相对含量（%）
硝态 N 化合物		methyl-acetyl-hydroxy-pyrrolidine-2-carboxylate	3.03	pyridine	3.44	1-methy-1-H-pyrrole	0.98
		4,5-dimethyl-4,5-dihydro-1H-pyrazole	1.10	1H-pyrrole	3.30	pyrrole	1.60
		1-(1H-pyrrlol-3-yl-) ethanone pyridin-3-ol	0.72	5-methyl-4,5-dihydro-1H-pyrazole	1.84	3-amino-4-methoxybenzoic acid	2.24
		7a-methyl-5,6,7,7a-trimethylcyclo-1H-inden-2(4H)-one	4.01	1H-pyrrole-2-carbaldehyde	1.99	2-(2-methylprop-1-enyl) cyclohexanone-oxime	3.08
		1-(2,3,5-trimethoxyphenyl) propan-2-amine	4.29	6-methoxypyridin-2-amine	1.37	5-methoxy-4-methylnicotinamide	2.17
			1.68	3-amino-4-methoxybenzoic acid	1.42	3,5-dinitrobenzamide	3.40
				5H-cyclopental[b]pyridine	5.27	1-(2,3,5-tromethoxyphenyl) propan-2-amine	2.20
				4-ethyl-piperdine	1.92	5,6,7,8-tetramethoxy-1,2,3,4-tetrahydroisoquinoline	2.05
				3-hydroxyphenylalanine	2.45		

续表

	主要组分	水花生 相对含量(%)	主要组分	水芹 相对含量(%)	主要组分	蒲草 相对含量(%)
总计		14.83		23.00		17.72
酚类	pyrocatechol	2.18	phenol	3.60	phenol	2.26
	4-ethyl-2-methoxyphenol	2.36	O-cresol	4.58	m-cresol	3.02
	resorcinol	2.19	2,3-dimethylphenol	1.85	2-methoxyphenol	4.77
	2,6-dimethoxyphenol	2.17	3-ethylphenol	2.87	pyrocatechol	3.08
	2,4-di-tert-butylphenol	4.54	pyrocatechol	1.61	2-methoxy-4-vinylphenol	2.09
	4-allyl-2,6-dimethoxyphenol	3.21	2-ethyl-6-methylphenol	2.01	2,4-dimethoxyphenol	4.24
			resorcinol	2.54		
总计		16.65		19.06		19.46
其他氧化物	1-hydroxypropan	9.71	1-hydroxypropan	5.53	1-hydroxypropan	9.63
	cyclohexane-1,2,4,5-tetraol	1.71	acetic acid	6.38	butane-1,3-diol	2.78
	acetic acid	5.16	cyclopentanone	1.23	acetic acid	5.17
	cyclopent-2-enone	2.84	4,5-dimethyl-4,5-dihydro-1H-pyrazole	1.21	cyclopentanone	1.11
	furan-2-carbaldehyde	5.30	cyclopent-2-enone	2.56	cyclopent-2-enone	1.71
	2-methylcyclopent-2-enone	1.77	furan-2-ylmethanol	2.53	2-methycyclopent-2-enone	1.88
	5-methylfuran-2-carbaldehyde	7.75	2-methycyclopent-2-enone	2.77	1-(furan-2-yl)ethanone	1.58

续表

水花生		水芹		蒲草	
主要组分	相对含量(%)	主要组分	相对含量(%)	主要组分	相对含量(%)
2-hydroxy-3-methylcyclopentanone	4.97	dihydrofuran-2(3H)-one	2.11	2,3-dimethycyclopent-2-enone	3.18
3,4-dimethyl-cyclopent-2-enone	2.22	3-methylcyclopent-2-enone	4.33	3-methycyclopent-2-enone	1.33
4,4-dimethylcyclohex-2-enone	4.95	3-methylcyclopentane-1,2-dione	3.58	3-methycyclopentane-1,2-dione	2.71
3-ethyl-2-hydroxycyclopent-enone	1.37	2,3-dimethylcyclopent-2-enone	2.33	3-ethyl-2-hydroxyclopent-2enone	1.26
4-(2-hydroxypropyl)-3,5,5-trimethyl cyclohex-3-enone	1.08	4,4-dimethylcyclohex-2-enone	4.24	5-hydroxy-4a-methyloctahydro naphthalen-1(2H)-one	1.27
2-methyl-1-(2,4,6-trimethoxyphenyl) propan-1-ol	1.68	3-ethyl-2-hydroxycyclo-2-enone	2.66	1,2-dimethoxy-4-(prop-1-enyl) benzene	5.43
2,3,6-trimethylnaphalene	5.47	3,6-dimethylcyclohex-(4-ene-1,2diyl) dimethanol	1.79	1,2,4-trimethoxy-5-(prop-1-enyl) benzene	1.88
		2-ethyl-1H-4 acid-carboxylic acid	1.33		
		2-(3,3-dimethylcyclohexylidene)ethanol	1.81		
总计	55.98		46.39		40.92

2.3.3
氮磷营养元素在热解过程中的迁移转化过程及机理解析

在热解过程中,氮磷元素可以随着挥发性组分的逸出分布在所有三相热解产物中。在本研究中,我们首先通过以下公式(式(2.23)~式(2.28))计算出氮磷元素在不同热解产物中的分布情况(TN = 氮元素分布情况;TP = 磷元素分布情况):

$$TN(生物炭) = \frac{生物炭中的氮含量 \times 生物炭的量}{生物料中的氮含量 \times 生物料的量} \times 100\% \quad (2.23)$$

$$TN(生物油) = \frac{生物油中的氮含量 \times 生物油的量}{生物料中的氮含量 \times 生物料的量} \times 100\% \quad (2.24)$$

$$TN(热解气) = 100\% - TN(生物炭) - TN(生物油) \quad (2.25)$$

$$TP(生物炭) = \frac{生物炭中的磷含量 \times 生物炭的量}{生物料中的磷含量 \times 生物料的量} \times 100\% \quad (2.26)$$

$$TP(生物油) = \frac{生物油中的磷含量 \times 生物油的量}{生物料中的磷含量 \times 生物料的量} \times 100\% \quad (2.27)$$

$$TP(热解气) = 100\% - TP(生物炭) - TP(生物油) \quad (2.28)$$

所得结果如图2.22所示,经过热解后,生物质中的氮元素主要还是集中在固相的热解生物炭中,并且热解温度越低,固体生物炭中的氮元素分布比例越大,不同湿地植物进料,其生物炭中氮的分布最大比例在60.5%~67.5%之间。而随着热解温度的升高,氮元素不断挥发至生物油或者气相组分中,导致固相残留的氮元素不断减少,而生物油和热解气中氮元素的分布随着热解温度的升高不断上升。其中,蒲草在热解温度为873 K时,产物中生物油中氮的分布将超过在生物炭中的分布。从环境角度而言,热解气中的氮元素主要以氮氧化合物及其前体(NH_3,HCN,NO等)形式存在[101],会对环境产生较为严重的危害,因此为了尽可能降低气相中的氮元素分布,热解温度不宜过高。和氮元素相比,磷的挥发性更低,因此其热解后在生物炭中的分布比例更高,如图2.22所示,磷在生物炭中的最大分布比例分别为85.4%(水芹,623 K)、90.7%(蒲草,623 K)和93.4%(水花生,673 K)。对于蒲草和水花生生物质而言,其热解过程中磷在生物炭中的分布随着温度的升高不断下降,相应地,生物油和热解气中磷的分布则不断上升。因此,低的热解温度有利于蒲草和水芹生物质中磷元素在生物炭中分布,但对水芹生物质中磷的分布影响不大。

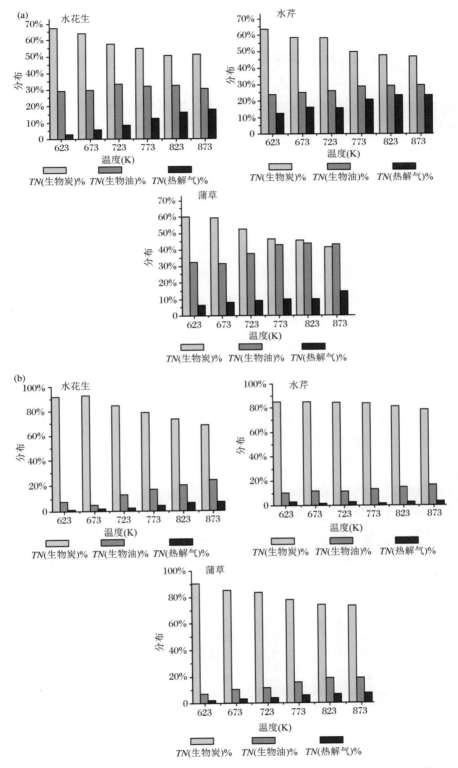

图 2.22　不同热解温度下,湿地植物生物质中氮磷元素在热解产物中的分布情况:
(a) TN;(b) TP

综上所述,热解温度是影响氮磷元素在热解产物中的分布情况的关键因素,低热解温度有利于氮磷在生物炭的分布,方便随后的提取回收过程。而在高的热解温度下,氮磷元素都有可能挥发至生物油和热解气中,增加其回收难度,并且可能引起新的环境问题。但是,热解温度过低会显著降低生物油的产率,也不利于资源的回收,因此,如何保证在获得最大生物油产率的情况下,同时使氮磷元素不至于大量挥发出来是一个需要解决的问题。在后续的研究中,我们将对这方面进行深入的研究,如考虑在热解过程加入催化剂,从而降低获得最佳生物油产率和品质的热解温度,或者加入固定剂将氮磷固定在生物炭中。

为了进一步分析热解过程中氮磷元素的转化过程,及其在生物炭中的主要存在形式,我们对生物炭中的氮磷进行了一系列浸出实验。生物炭中铵态氮通过 $1\ mol\cdot L^{-1}$ 的氯化钾溶液浸出,而硝态氮则通过(质量比为2%)的硫酸钾溶液浸出,剩余的氮元素通过凯氏定氮法消解测定。结果如图2.23所示,在生物质热解前,只有不到17%的氮元素可以通过两个浸出溶液提取出来,表明生物质本身离子态的氮元素含量很低。而热解过后,在生物炭中,无论是铵态氮还是硝态氮,其含量都有增加,其中铵态氮的增加最为明显。在这些可以被浸出的氮元素中,铵态氮的比例远高于硝态氮,占80%以上,这主要是因为热解过程是在无氧气氛中进行的,不利于硝态氮的生成,而湿地植物生物质中的氮元素(氨基酸、生物碱、叶绿素等)在热解条件下发生降解,可以形成小分子的含氮物质或者无机铵态氮等[102],从而保留在生物炭中。

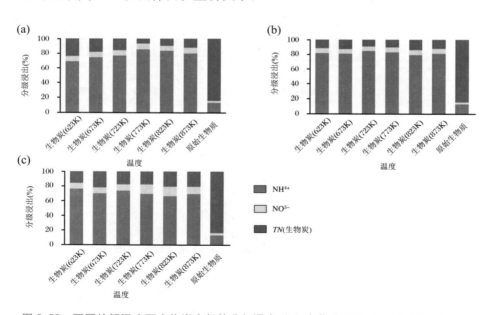

图2.23 不同热解温度下生物炭中氮的分级浸出:(a) 水花生;(b) 水芹;(c) 蒲草

生物炭中磷的分级提取是通过碳酸氢钠（$NaHCO_3$）溶液（pH = 8.5）和盐酸/氟化铵（HCl/NH_4F，0.025 mol·L^{-1}）来实现的，其中 $NaHCO_3$ 溶液提取的磷是易溶解的磷（Diffluent Phosphate，DP），这部分磷被认为是可被植物迅速利用的磷，而 HCl/NH_4F 溶液提取的磷是和铁、铝、钙等金属结合的磷（Binding Phosphate，BP），这部分磷也是可以被植物利用的磷，此外，还有部分磷不能被这两种溶液提取出来，可以在完全消解之后再进行测定。磷的提取结果如图 2.24 所示，在所有的不同热解温度下获得的生物炭样品中，磷的平均提取率为 57%，比未热解生物质中磷的提取率稍高（平均为 42%），这主要是因为在生物质中，磷主要是以有机态的形式结合在多糖或者其他成分中，比较难以通过无机溶液提取出来，而经过热解之后，有机结合的磷发生键的断裂，形成无机态的磷存在于生物炭中，可以通过无机溶液加以提取。此外相比于原始生物质，热解的生物炭的体积和质量都大为降低，从而可以显著减少浸出溶液的消耗和废弃物的排放。

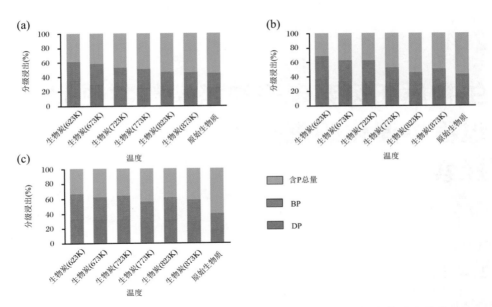

图 2.24　不同热解温度下生物炭中磷的分级浸出：(a) 水花生；(b) 水芹；(c) 蒲草

在本节研究中，我们选择了三种代表性的湿地植物进行快速热解，研究其热解过程的影响因素及氮磷营养元素的迁移转化和分布及机理，根据实验结果，可以获得以下结论：

（1）湿地植物生物质具有很强的氮磷营养元素富集能力，其氮磷含量较普通木质纤维类生物质大为提高。快速热解作为一个常用的废弃生物质资源化技术，对湿地植物资源化利用及氮磷回收具有很好的应用价值。

（2）热解温度是影响湿地植物热解产物产率、组成及性质的最主要的因素，同时影响着氮磷元素在热解产物中的分布。热解温度的升高不利于氮磷在生物炭中的富集，但是温度过低也会影响生物油的产率和性质，因此，选择合适的热解温度对于湿地植物的热解至关重要。

(3) 热解所得生物油的产率在 40% 以上，其主要成分包括酚类、醛酮类、羧酸类以及一些含氮杂环或者胺类化合物。尽管其热值不高，不适合直接用作燃料使用，但是因其成分复杂，可以用作提取高附加值化学品的原料，具有较高的经济前景。

(4) 在热解过程中，大量的氮磷元素从有机态转化为无机离子态或者结合态，仍然集中在生物炭中，可以通过无机盐溶液浸出的方法分级提取，氮元素和磷元素的平均回收率分别为 76%、57%。

2.4
生物质和废弃电子垃圾塑料共热解过程中溴化阻燃剂的分解转化机理探索

2.4.1
概述

自 20 世纪 40 年代开始的塑料工业化以来，塑料废弃物的产生速率不断加快[103]，其中增长最快的是废弃电子电气垃圾（Waste Electrical and Electronic Equipments，WEEEs）塑料[104-111]。根据 2009 年的估算，全球每年产生的 WEEEs 塑料超过 4 亿 t，比其他市政废弃物的增长速率快 2 倍。常规的处理 WEEEs 塑料的方法包括填埋、机械回收和焚烧[112-114]。然而，这些方法都存在或多或少的不足之处。例如，WEEEs 塑料的填埋通常需要占用很大的填埋土地，此外，填埋产生的垃圾渗滤液含有各种污染物（重金属、持续性有机物），会对地下水和周围土地产生严重污染,进而威胁人类的健康。机械回收方法需要高的劳动成本,因为在机械处理之前需要对WEEEs进行人工分离和归类。而在WEEEs塑料的焚烧过程中,尽管可以产生大量的能量用于发电或者其他用途,但是燃烧过程中产生类二噁英化合物或者颗粒污染物是不可避免的,这会造成严重的环境问题[115]。更重要的是,WEEEs塑料中的溴化阻燃剂(Bromimated Flame Retardants, BFRs)是一种广泛用作降低塑料着火危险的试剂[116],存在潜在的环境危害。当WEEEs塑料被填埋时,几乎所有的BFRs都是难降解的,并且容易产生生物积累的污染物,而当机械处理或者焚烧WEEEs塑料时,BFRs容易分解氧化产生高毒性的持久性有机污染物,例如多溴联苯并二噁英或者呋喃类(PBDD/Fs)[117]。为了避免BFRs可能存在的环境风险,迫切需要寻找处理含有BFRs的废弃塑料的新方法。

快速热解作为一种在缺氧环境下热化学降解有机物的方法,被认为是一种环境友好且经济可行的WEEEs塑料回收方法[118-125]。其主要有如下几个优势: ① 由于快速热解过程是在缺氧条件下进行的,因此过程中生成多溴联苯并二噁英或者呋喃类污染物的机会将大为降低;② 快速热解过程的能量消耗很低(通常只有WEEEs塑料所含能量的10%左右)[126];③ 在热解过程中,塑料中的聚合物结构发生热分解,转变成一些小分子气体或者液体,可以用作燃料或者是化学品[118,127]。最近很多研究都已经证实快速热解用于回收WEEEs塑料在技术上是可行的[127-129]。然而,在WEEEs塑料的热解过程中,热解油的产率和性质通常受到WEEEs的低氢碳比组成的直接限制。如果在热解过程中,提供一种高氢碳比的原料用作氢的来源提供给废弃塑料的热解,可以显著提高热解油的产率和性能(热值、稳定性等)。木质纤维类生物质含有较高的碳氢比,可以和WEEEs塑料进行共热解,作为氢供体提高热解油产率和稳定性。在热解过程中,除了回收液体和气体燃料资源外,BFRs的迁移转化是一个重要的问题。尽管在焚烧和机械处理过程,BFRs的分解转化过程的研究已经大量见于文献报道,然而热解过程是在无氧环境中进行的,BFRs经历的分解转化行为和焚烧和机械处理过程将大为不同,需要深入研究。此外,BFRs在热解过程中的分解转化行为的阐明也将为我们在接下来的工作中如何避免在热化学处理WEEEs塑料时产生PBDD/Fs,同时选择性的将其中的溴元素富集在热解生物炭中提供科学依据。

在本节研究中,我们采用木屑生物质和WEEEs塑料进行共热解,获得热解油,并研究其中的BFRs的分解转化过程,利用热重-红外光谱-质谱(TG-FTIR-MS)对WEEEs塑料的分解过程进行在线监测和分析,解析其热分解过程的机理。

2.4.2
生物质与废弃电子塑料的基本特性分析

生物质(木屑)采自本地的一个木材处理厂,其基本元素(碳、氢、氧、氮)组成通过元素分析仪进行测定,生物质和 WEEEs 塑料的矿质元素通过化学消解后采用 ICP-AES 进行测定,它们的热稳定性使用热重分析仪进行表征(DTG-60H/DSC-60,Shimadzu Co.,Japan)。WEEEs 塑料样品是从当地的一个废弃塑料回收厂收集的。在使用之前,WEEEs 塑料先通过高速粉碎仪进行破碎,收集其中粒径小于 0.12 mm 的颗粒作为热解实验进料。WEEEs 塑料中的溴化阻燃剂成分为十溴二苯乙烷,是一种稳定的、具有很好阻燃效果的化合物,目前被大量用作十溴二苯醚的替代品[130]。表 2.7 列出了十溴二苯乙烷的基本性质。

表 2.7 十溴二苯乙烷的基本性质

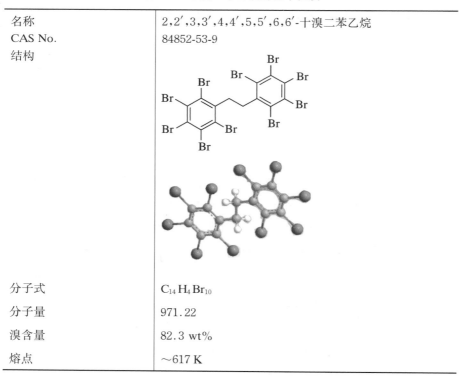

名称	2,2′,3,3′,4,4′,5,5′,6,6′-十溴二苯乙烷
CAS No.	84852-53-9
结构	
分子式	$C_{14}H_4Br_{10}$
分子量	971.22
溴含量	82.3 wt%
熔点	~617 K

表 2.8 列出了生物质和 WEEEs 塑料的基本组成和性质。由表可以看出,WEEEs 塑料的氢碳比(H/C)和氧碳比(O/C)分别为 0.11 和 0.90,远远低于生物质中的(其 H/C 和 O/C 分别为 0.75 和 1.56),这表明 WEEEs 塑料有比生物质更高的能量密度[131]。和生物质相比,WEEEs 中有两种新的大量存在的元素

被检测到,其中溴元素是 BFRs 的主要成分,而锑元素是阻燃助剂 Sb_2O_3 的主要成分。

表 2.8　生物质和 WEEEs 塑料的基本组成和性质

项目	木屑生物质	WEEEs 塑料
近似分析(wt%)①		
水分	3.32	0.07
挥发性物质	79.40	92.48
固定碳	15.65	5.12
灰分	1.63	2.33
元素分析（wt%）②		
C	46.52	80.20
H	6.03	6.78
N	0.20	0.13
O	46.81	8.50
Br	—	4.39
H/C（摩尔比）	1.56	0.90
O/C（摩尔比）	0.75	0.11
热值(MJ·kg^{-1})③	18.43	37.57
矿质元素分析（mg·kg^{-1}）④		
K	392	24
Na	316	158
Ca	1678	3654
Mg	376	606
Al	68	448
Fe	98	216
Sb	—	1420

注:① 干燥样品。

② 干燥无灰分样品。

③ 热值(MJ·kg^{-1}) = $(3.55 C^2 - 232 C - 2230 H + 51.2 C \times H + 131 N + 20600)$/1000,见公式(2.17)。

④ 干燥样品。

生物质和 WEEEs 塑料的热稳定性通过热重分析仪来表征,结果如图 2.25 所示。对于 WEEE 塑料,其最初的失重开始于 300 ℃左右,在其差热曲线上出现两个降解峰。其中第一个峰出现在 300~368 ℃之间,可以归因为 BFRs 和低聚的塑料分子降解。另一个峰出现在370～495 ℃之间，被认为是那些具有高聚合度的塑料分子的分解。对于木屑生物质，其最初的失重开始于100 ℃，是由水分的挥发导致的。主要的失重过程发生在200～400 ℃之间，这是由生物质中的纤维素、半纤维素和木质素的大量分解挥发导致的。

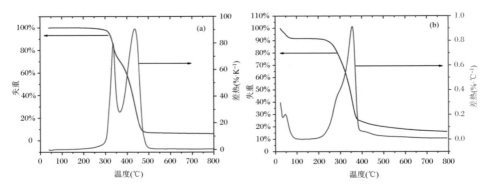

图2.25　WEEEs塑料(a)和木屑生物质(b)的热重及差热曲线

2.4.3
生物质和WEEEs塑料的共热解过程及影响因素

热解之前，木屑和WEEEs塑料先按照不同的比例均匀地混合作为热解进料，热解的基本操作和2.2.4描述基本一致，不同的是在热解过程中产生的不可冷凝组分将通过两个吸收瓶进行吸收，其中一个装有100 mL丙酮，另一个装有100 mL NaOH溶液(1 mol·L^{-1})。

热解前后WEEEs塑料中溴的化学状态变化通过X射线光电子能谱(XPS，ESCALAB250，Thermo-VG Scientific，UK)进行分析。热解前后溴在进料和产物中的含量是在碱性消解之后通过ICP-MS（Plasma Quad 3，Thermo-VG Elemental.，UK)进行测定的。其具体步骤如下：

(1) 对生物炭中的溴进行测定。0.10 g的样品和0.80 g的烧结试剂(45 g的Na_2CO_3和30 g的ZnO混合制成)均匀混合，然后再将另外0.80 g的烧结试剂覆盖在其表面，最后将其置于马弗炉中，在550 ℃下灼烧1 h，冷却后将所得残渣用氢氧化钠溶液(1 mol·L^{-1})溶解，再用盐酸溶液调节pH至1.0并定容至100 mL，最后使用ICP-MS测定其中溴元素的浓度，计算出溴在生物炭中的含量。热解油和气体吸收瓶中溴的含量测定和生物炭的测定方法基本相同，唯一不同的是在加烧结剂前，先用0.5 mL的氢氧化钠溶液将热解油或者吸收瓶中的液体分散。

（2）WEEEs 塑料的热解过程通过 TG-FTIR-MS 进行在线监测。实验过程如下：将 5 mg 的样品置于样品槽中，分别以 10 ℃·min^{-1}，50 ℃·min^{-1}，200 ℃·min^{-1} 的升温速率，在 100 mL·min^{-1} 的氦气载气条件下从室温升高至 800 ℃，热解过程中产生的挥发性物质通过红外光谱和质谱在线分析其含量和化学结构。TG-FTIR-MS 之间的连接系统也一直加热避免挥发性物质的冷凝。FTIR 扫描波数在 4000～400 cm^{-1} 之间，每隔 2.5 s 采集一次数据。MS 扫描在 70 eV 条件下进行，主要检测一些被认为是 WEEEs 塑料和 BFRs 分解产物的离子峰（荷质比小于 200）。那些荷质比大于 200 的离子峰在质谱无法直接准确检测出来，因此，我们先采用在线气相色谱分离，再使用质谱进行检测，这个方法可以检测到荷质比在 500 以内的离子峰。热解之后，获得的生物炭的形貌通过扫描电子显微镜（SEM,Sirion 200,FEI electron optics company,USA）进行观测，溴的化学状态变化通过 XPS 进行分析。

由于 WEEEs 塑料和生物质的特性和组成大为不同，因此它们的共热解是一个非常复杂的过程。实验条件（热解温度、WEEEs 塑料和生物质的混合比等）不同，热解产物（热解油、生物炭和气体）的产率和性质都有明显区别[132]。图 2.26 所示的是不同热解温度下热解产物的产率分布。从中可知热解生物炭的产率从 400 ℃ 时的 41.7% 迅速降低至 600 ℃ 时的 14.3%，而热解气的产率则从 10% 上升至接近 30%。这种现象可以解释如下：根据 TGA 分析，在 400 ℃ 时，WEEEs 塑料和生物质都已经开始分解了，其分子中的化学键（C—Br，C—C，C—O）断裂生成小分子物质挥发出去，并且随着温度的升高，挥发过程将会进行得更加迅速，导致了固体生物炭的产率不断降低。而随着温度的进一步升高，挥发组分将发生二次裂解，产生不冷凝气体组分，从而导致热解气体的产率不断增加。对于热解油，其产率从 400 ℃ 时的 48.5% 迅速上升至 500 ℃ 时的 62.3%，然后又下降到 600 ℃ 时的 55.1%。这可能是温度在 400～500 ℃ 区间时，生物质和 WEEEs 塑料的分解过程占主导，温度的升高有利于它们降解产生挥发性的组分，冷凝成热解油，使得其产率不断上升，而当温度在 500 ℃ 以上时，挥发性组分的进一步裂解产生小分子不可冷凝其他的过程将占主导地位，使得热解油的产率降低。相似的现象在其他相关文献中也有报道[67,133]。

除了温度之外，生物质与 WEEEs 塑料的混合比也是一个影响热解产物分布和性质的重要因素。图 2.27 所示是不同混合比条件下热解产物的产率分布。从中可知，当生物质的混合比从 33% 增加到 50% 时，热解油的产率从 55.0% 升高至 62.3%，随着生物质的混合比例进一步增加到 80%，热解油的产率稍微下

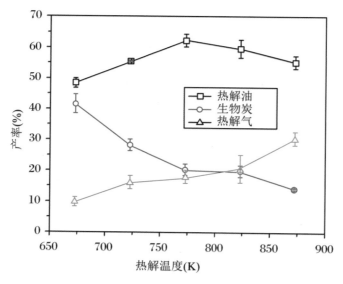

图 2.26　不同温度下 WEEEs 塑料和木屑生物质共热解产物的
产率分布情况（WEEEs 塑料和生物质混合比为 1∶1）

降到了 59.0%。和分别单独热解生物质和 WEEEs 相比，共热解这两种进料的热解油产率大大提高了（单独热解生物质的油产率为 46.3%，而单独热解 WEEEs 塑料的油产率为 53.1%），表明生物质和 WEEEs 塑料在共热解过程中确实存在着协同作用。其主要机理可以分析如下：如表 2.8 所示，一方面，生物质的氢碳比远远高于 WEEEs 塑料的碳氢比，在热解过程中可以作为氢的供体，促进 WEEEs 塑料的分解[134]，从而提高热解油的产率。另一方面，生物质热解过程产生的水也可以作为一种活性物质促进 WEEEs 塑料热解生物炭的进一步分解产生挥发性物质，从而增加热解油的产率[93,135]。

作为一个易挥发性的元素，溴在共热解过程中可以在气、固、液三相中都有分布，而溴元素对热解油作为燃料使用来说是有害的，因此，研究溴元素在热解过程中的分布情况及影响因素对如何控制溴元素在热解油的含量具有重要的指导意义。溴元素在热解油、生物炭和热解气体中的分布情况可以通过以下公式进行计算：

$$\mathrm{Br}(生物炭)(\%) = \frac{生物炭中的溴含量 \times 生物炭的产量}{生物质进料中的溴含量 \times 生物质进料的质量} \times 100\% \tag{2.29}$$

$$\mathrm{Br}(解热油)(\%) = \frac{生物油中的溴含量 \times 生物油的产量}{生物质进料中的溴含量 \times 生物质进料的质量} \times 100\% \tag{2.30}$$

$$\mathrm{Br}(热解气)(\%) = \mathrm{Br}(丙酮吸收液)(\%) + \mathrm{Br}(\mathrm{NaOH}吸收液)(\%)$$

$$= \frac{\text{丙酮吸收剂中的溴含量} \times \text{丙酮的总量}}{\text{生物质进料中的溴含量} \times \text{生物质进料的质量}} \times 100\%$$

$$+ \frac{\text{NaOH 溶液中的溴含量} \times \text{NaOH 溶液的总量}}{\text{生物质进料中的溴含量} \times \text{生物质进料的质量}} \times 100\%$$

(2.31)

$$\text{总的 Br 回收率(\%)} = \text{Br(生物炭)\%} + \text{Br(热解油)\%} + \text{Br(热解气)\%} \quad (2.32)$$

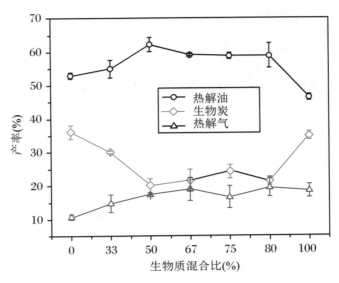

图 2.27　不同 WEEEs 塑料和木屑生物质混合比条件下的产率分布情况

热解温度为 500 ℃

不同温度条件下,溴元素在热解产物中的分布情况以及总的溴回收率如图 2.28(a)所示。结果显示,在低的热解温度下,溴元素主要分布在热解油中,而随着温度的升高,溴元素在热解生物炭中的分布比例不断上升。这个现象和其他挥发元素的热解行为大为不同,例如氮磷元素在热解生物炭中的分布都是随着温度的升高而不断降低的[136-137]。这种不寻常的现象可以解释如下:热解产物中的溴元素主要来源于十溴二苯乙烷,这是一种高度稳定的化合物,其热化学变化如图 2.28(b)所示,主要分解温度在 400~600 ℃ 之间。因此在低的热解温度下,十溴二苯乙烷主要是直接以分子的形式挥发出来,转移至热解油组分,而随着温度的升高,十溴二苯乙烷发生分解,产生大量的活性溴化物(HBr 以及溴自由基等)[138-139],可以被热解生物炭中的无机成分和碳骨架结构捕获而存在其中。

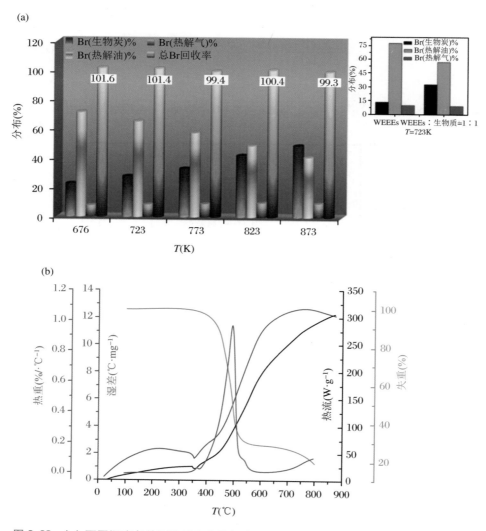

图 2.28 （a）不同温度条件下溴元素在热解产物中的分布情况；（b）十溴二苯乙烷的热化学特性分析

图(a)中生物质和 WEEEs 塑料混合比为 1∶1（右上插图）以及单独热解和混合共热解条件下溴元素的回收率比较

2.4.4

热解过程中溴化阻燃剂的分解转化过程和机理

为了对热解过程中 BFRs 的分解机理有更深的了解，我们采用 TG-FTIR-MS 对热解过程进行在线监测，通过对其热解行为的分析以及热解产物的结构和含量的表征来阐述 BFRs 的分解机理。图 2.29 所示的是不同升温速率条件下 WEEEs 塑料的热重-差热曲线。如图所示,WEEEs 塑料的热解过程分成两个阶段:第一阶段发生在 280～380 ℃ 之间,失重为 23%,可以认为是塑料中的十溴二苯乙烷的挥发和分解过程。第二阶段失重发生在 380～550 ℃ 之间,对应于塑料中的高分子链的分解过程产生小分子挥发性物质。在不同的升温速率下,差热曲线(失重速率)存在两个峰值,其中第一个峰值分别发生在 317 ℃,355 ℃ 和 374 ℃(对应于升温速率 10 ℃·min^{-1},50 ℃·min^{-1} 和 200 ℃·min^{-1}),而第二个峰值分别发生在 422 ℃,457 ℃ 和 486 ℃(对应于升温速率 10 ℃/min,50 ℃/min 和 200 ℃/min)。

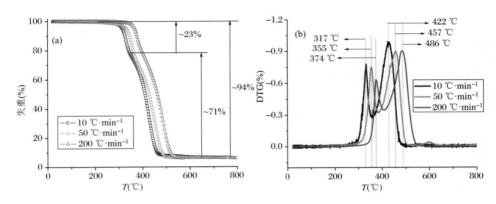

图 2.29 (a) 不同升温速率下 WEEEs 塑料的热重曲线;(b) 不同升温速率下 WEEEs 塑料的差热曲线

接下来我们对在差热曲线峰值情况下挥发出来的组分进行了在线 GC-MS 分析。图 2.30(a) 所示是在升温速率为 200 ℃·min^{-1}、温度为 374 ℃(峰值 1) 条件下挥发组分的 GC-MS 图谱。在 GC 谱图中有 4 个主要的峰,MS 分析其结构分别为苯、甲苯、苯乙烯和 2-甲基苯乙烯,这些结构都是塑料高分子链分解的主要产物[138,140]。还有一些具有不止一个苯环的芳香化合物被检测到了,这些化合物被认为是高分子链分解的主要产物进一步反应或者是链分解不完全的产物。除此之外,还发现了 7 个溴化的芳香化合物在 GC-MS 图谱中,这些化合物要么是由溴化阻燃剂直接分解产生的(GC 峰 5,6,7),要么是由活性溴组分(溴自由基和溴化氢)和塑料高分子链分解产物进一步反应产生的(溴苯和 GC 峰 1,2,3)[141]。图 2.30(b) 所示是在升温速率为 200 ℃·min^{-1}、温度为 486 ℃ 条件下(峰值 2) 产生的挥发组分的 GC-MS 图谱。值得注意的是,和在 374 ℃ 下鉴定的化合物相比,在 486 ℃ 下发现的溴化物都只有一个苯环,可以认为是溴化阻燃剂分解碎片进一步分解产生的或者是活性溴组分与其他芳香化合物反应的产物。除此之外,在 GC-MS 图谱中也发现了很多的芳香化合物,包括 5 个单芳香化合物和 5 个多环芳烃。

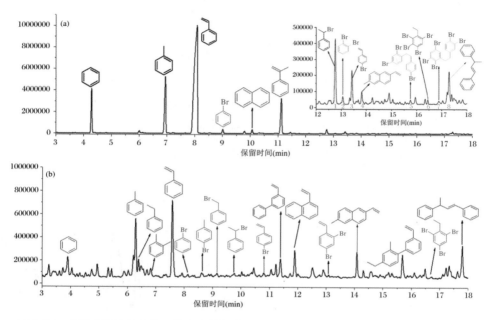

图 2.30　差热曲线峰值条件下的产生的挥发组分的 GC-MS 图谱分析：(a) 374 ℃；(b) 486 ℃ 升温速率为 200 ℃·min^{-1}

除了 GC-MS 可以分析到的那些芳香化合物，一些小分子化合物可以通过 TG-FTIR 来检测。在研究中，我们选择检测 3 种主要的挥发气体组分（HBr，CH_3Br 和 CH_4），它们的分子结构简单，只有一个特征的红外吸收峰，分别在波数为 2450 cm^{-1}，954 cm^{-1} 和 3028 cm^{-1} 处[142-144]，结果如图 2.31 所示。这 3 种化合物主要形成在 300~650 ℃ 之间，和热重曲线中主要失重区间是一致的。这 3 种化合物的释放都存在两个峰值，和差热曲线的峰值是类似的。但是这 3 种化合物的释放曲线中的温度值比差热曲线中的温度值要高 15~35 ℃，可能的解释是这 3 种化合物并不是 WEEEs 塑料直接降解的产物，而是最初降解产物进一步反应的产物。

升温速率对热解产物的产率和性质有重大影响[145]。如图 2.31 所示，在低的升温速率（如 10 ℃·min^{-1} 和 50 ℃·min^{-1}）下，HBr，CH_3Br，和 CH_4 的释放都不太明显，但是其释放趋势在高的升温速率（200 ℃·min^{-1}）条件下大为增强。CH_3Br 和 CH_4 在 200 ℃·min^{-1} 时的释放曲线具有类似的趋势，都有两个释放峰值，分别发生在 420 ℃ 和 510 ℃ 左右。而对于 HBr 而言，尽管其在 200 ℃·min^{-1} 时的释放速率明显大于在 10 ℃·min^{-1} 和 50 ℃·min^{-1}，但是在低的升温速率下，其释放的温度范围在 300~800 ℃ 之间，明显比在 200 ℃·min^{-1} 时的温度区间（490~580 ℃）大了很多。这主要是因为 HBr 是一个高极性的分子,容易被热解连接系统的表面所吸附,而吸附过的 HBr 的在释放速率随着温度的升高而不断增加[141]。

图 2.31(d)所示的是在 200 ℃ · min^{-1} 的升温速率下 WEEEs 塑料所有热解产物的 3D FTIR 图谱。从中可知,FTIR 吸收峰的温度依赖性和差热曲线是一致的:热解产物的红外吸收在 490 ℃ 时达到最大值,这也是热解过程中 WEEEs 塑料失重最快的时候(见图 2.29(b))。热解产物的不同红外吸收峰鉴定列于表 2.9。从中可知,溴化烃的存在可以通过波数在 931 cm^{-1},954 cm^{-1} 和 1054 cm^{-1} 红外吸收峰证实(分别是亚甲基、甲基和芳香碳-溴键的伸缩振动)[146-147]。波数在 2800~3100 cm^{-1} 和 1600~2000 cm^{-1} 之间的红外吸收峰分别是 C—H 和 C—C (C=C)伸缩振动[148]。值得注意的是,根据以上结果分析,在热解过程中只有烃和溴代烃的生成,而没有 PBDD/Fs 高毒性持久性有机污染物的形成,因为在 FTIR 的图谱中没有观测到 O—H 和 C—O 基团的振动。

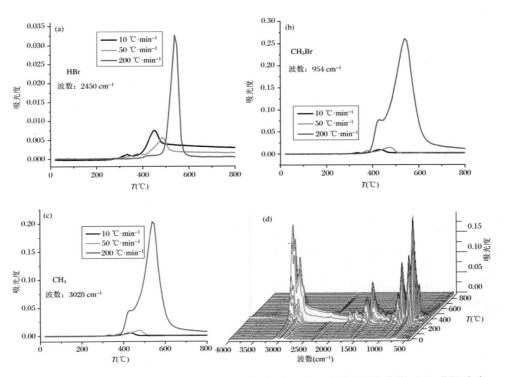

图 2.31 (a~c) 不同升温速率条件下 HBr,CH$_3$Br 和 CH$_4$ 的 FTIR 吸收曲线;(d) 升温速率为 200 ℃ · min^{-1} 条件下所有热解产物的 3D FTIR 图谱

表 2.9　热解产物的不同红外吸收峰鉴定

功能基团	波数位置（cm^{-1}）
芳香 C—H 伸缩振动	3093±3
末端乙烯基 C—H 伸缩振动	3082±4
甲烷 C—H 伸缩振动	3028
内部烯基 C—H 伸缩振动	3009±2
甲基 C—H 伸缩振动	2961±2,2870±3
亚甲基 C—H 伸缩振动	2920±4,2858±5
H-Br 伸缩振动	2450
芳环组合谱带	1944±4,1877±2,1804±5,1752±3,1691±1
烯基 C=C 伸缩振动	1630±7
芳环 C=C 伸缩振动	1601±3,1576±4,1495±3,1450±5
乙烯基 C—H 平面内弯曲振动	1412±6
亚乙烯基 C—H 平面内弯曲振动	1290±8
芳香基 C—H 平面内弯曲振动	1202±3,1083±5,1028±3,1018±4
芳香 C—Br 伸缩振动	1049±6
甲基 C—Br 伸缩振动	954
亚甲基 C—Br 伸缩振动	931±11
乙烯基 C—H 平面外弯曲振动	989±6,907±4
芳香基 C—H 平面外弯曲振动	840±5,774±3,755±6,748±4,731±4
芳环 C—H 弯曲振动	694±7

尽管 TG-FTIR 可以提供很多关于热解过程中挥发性产物的信息，但是这种方法仍然存在一些缺点。例如 TG-FTIR 不可以给出热解产物的具体分子结构。此外，热解过程中会产生较强的 CO_2 红外吸收信号，会对其他化合物的红外信号产生较大的干扰。为了更进一步分析 BFRs 的分解机理，我们对热解产物进行了 TG-MS 分析，确定其分解产物的结构和挥发行为。9 种主要代表性的分解产物，包括 4 种丰度最高的芳香烃类和 5 种溴代芳烃类化合物在不同升温速率条件下的释放行为通过 TG-MS 来监测。对于 4 种芳香化合物（苯、甲苯、苯乙烯和 2-甲基苯乙烯，见图 2.32(a~d)），它们在升温速率为 200 ℃·min^{-1} 时的释放远高于在低的升温速率（10 ℃·min^{-1} 和 50 ℃·min^{-1}）条件下的，这个结果和 FTIR 的结果是一致的。这 4 种芳香烃类在 10 ℃·min^{-1} 和 50 ℃·min^{-1} 时的释放曲线含有两个峰值，分别出现在 420 ℃ 和 510 ℃，而当升温速率达到 200 ℃·min^{-1} 时，它们的释放只是在 510 ℃ 时出现一个峰值，表明不同的

升温速率可以导致 WEEEs 塑料产生的不同分解行为。在这 4 种芳香烃之中，苯和苯乙烯是塑料直接降解的终产物[149]，表现出最高的质谱丰度，分别达到 17×10^6 和 31×10^6。众所周知，苯、苯乙烯和甲苯都是重要的化工原料，可以用来生产许多精细化学品（例如医药、杀虫剂和染料等）[150-152]，而通过热解 WEEEs 塑料可以获得大量的这些化学原料，证明了热解是一个很好的回收利用 WEEEs 塑料的方法。

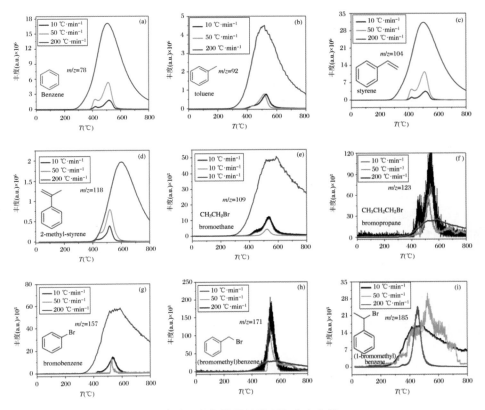

图 2.32　不同升温速率下热解过程产物的释放的 MS 响应曲线

和芳香烃类相比，溴代烃类的释放行为就复杂多了。如图 2.32(e) 和图 2.32(g) 所示，溴甲烷和溴苯的释放趋势与芳香烃类是相似的，都是在 $200\ ℃·min^{-1}$ 时表现出最高的释放速率，表明这两种溴代烃的生成方式可能和普通芳香烃类是一致的。而溴丙烷和溴甲基苯的释放行为则大为不同，它们的最大释放速率发生在 $10\ ℃·min^{-1}$ 时，暗示了这两种化合物经过了一个与众不同的形成途径，具体经过何种途径形成需要进一步探索。

根据 TG-FTIR-MS 的结果，我们可以对 WEEEs 塑料的热解行为和其中 BFRs 的分解过程有较为深入的了解。下面我们研究热解过程的动力学。WEEEs 塑料的热解动力学可以通过以下方程来表示：

$$r = \frac{d\alpha}{dt} = kf(\alpha) \tag{2.33}$$

$$k = Ae^{-E/(RT)} \tag{2.34}$$

其中，r 是反应速率，t 是反应时间，k 是动力学常数，A 是频率因子，E 是活化能，R 是气体常数，T 是热解温度，$f(\alpha)$ 是一个描述反应程度和速率的动力学方程，α 是热解过程中的失重率，可以由以下方程进行描述：

$$\alpha = \frac{w_i - w_t}{w_i - w_f} \tag{2.35}$$

其中，w_i，w_t 和 w_f 分别是样品的初始质量、在热解时间为 t 时的质量和最终质量。对于常规的固体分解反应[149]，$f(\alpha)$ 可以由下面的方程描述：

$$f(\alpha) = (1-\alpha)^n \tag{2.36}$$

其中，n 是和反应物消耗速率相关的反应级数。在一个升温速率恒定的热解反应中，n 的值可以固定为 1[153-154]。因此，WEEEs 塑料的热解动力学方程可以最终由以下方程来定义：

$$r = \frac{d\alpha}{dt} = kf(\alpha) = 1 - \alpha \tag{2.37}$$

因为热解过程是在恒定的升温速率下进行的，所以温度可以表示成反应时间的线性函数：

$$T = \beta t + T_i \tag{2.38}$$

其中，β 是升温速率，T_i 是初始温度。对方程(2.38)进行微分得到以下方程：

$$dT = \beta dt \tag{2.39}$$

根据方程(2.39)，方程(2.37)可以变形为 α 和 T 的函数：

$$\frac{d\alpha}{1-\alpha} = \frac{k}{\beta}dT = \frac{Ae^{-E/RT}}{\beta}dT \tag{2.40}$$

对方程(2.40)进行积分获得如下方程：

$$-\ln(1-\alpha) = \frac{A}{\beta}\int_0^t e^{\frac{-E}{RT}}dT \tag{2.41}$$

进一步对方程(2.41)采用修正 Caots-Redfern 方法[155]进行积分获得如下方程：

$$\ln\left[\frac{-\ln(1-\alpha)}{T^2}\right] = \ln\left[\frac{AR}{\beta E}\left(1 - \frac{2RT}{E}\right)\right] - \frac{E}{RT} \tag{2.42}$$

由于 $2RT/E$ 的值远小于 1，因此可以忽略不计。基于这个原因，方程(2.42)可以简化为

$$\ln\left[\frac{-\ln(1-\alpha)}{T^2}\right] = \ln\left[\frac{AR}{\beta E}\right] - \frac{E}{RT} \tag{2.43}$$

根据方程(2.43)，对 $\ln\left[\frac{-\ln(1-\alpha)}{T^2}\right]$ 和 $1/T$ 作线性图，反应活化能和频率

因子就可以分别通过这条直线的截距和斜率计算出来了,结果如表 2.10 所示。根据差热曲线,WEEEs 塑料的热解动力学可以分为两个阶段,第一阶段的反应活化能为 19.4 kJ·mol^{-1}(升温速率为 10 ℃·min^{-1}),远远低于第二阶段(73.3 kJ·mol^{-1})。同时,对于不同的升温速率,反应活化能的变化不大(第一阶段从 19.4 kJ·mol^{-1} 到 21.7 kJ·mol^{-1},而第二阶段从 73.3 kJ·mol^{-1} 到 75.2 kJ·mol^{-1})。类似的结果也见于不少文献报道[149,154]。这种现象可以解释为在惰性气体氛围下热解,WEEEs 塑料的分解过程主要涉及中间组分的生成以及其进一步分解产生小分子挥发产物的过程[156]。

表 2.10　不同升温速率条件下 WEEEs 塑料的热解动力学参数

升温速率 (K·min^{-1})	反应阶段	温度范围 (K)	反应活化能 (kJ·mol^{-1})	频率因子 (s^{-1})	相关系数 R^2
10	第一阶段	293~590	19.4	0.76	0.92
	第二阶段	590~748	73.3	70.3	0.96
50	第一阶段	293~628	21.3	1.52	0.89
	第二阶段	628~791	74.8	71.7	0.97
200	第一阶段	293~647	21.7	1.79	0.78
	第二阶段	647~818	75.2	72.1	0.97

图 2.33(a)所示的是 WEEEs 热解过程中溴化阻燃剂的分解机理示意图。对于十溴二苯乙烷,脱溴反应是其在热解过程中发生的最基本的反应,产生了溴原子小于十个的多溴二苯化合物(S1)以及大量的高反应活性的溴自由基[157]。化合物 S1 在高温的环境下将会进一步分解产生溴化的单芳香环化合物和更多的溴自由基。如 GC-MS 图谱证实的那样,在热解温度为 374 ℃时产生的化合物含有大量的多溴二苯化合物,而在 486 ℃时,所有的溴代烃都是单芳香环的化合物了。对于塑料聚合物本身,其聚合物链在热解过程中也会发生分解,产生大量的单芳香化合物(如苯、甲苯、苯乙烯等)和活性自由基(如·H,·CH$_3$,·CH$_2$CH$_3$)。这些活性自由基可以捕集溴自由基,形成很多其他的溴化合物(如 HBr,CH$_3$Br,CH$_2$CH$_3$Br 和一些溴化单环芳香烃)。除了有机组分可以捕获溴自由基之外,热解生物炭中的大量无机成分也可以捕获那些溴自由基形成无机溴化物而存在热解生物炭中。XPS 结果如图 2.33(b)所示,在热解之前,塑料中的溴的 XPS 图谱含有两个峰出现在结合能为 71.1 eV 和 72.0 eV 的位置,可以被认为是溴元素分别和 sp2 或者 sp3 碳原子以共价键结合[158]。而在热解之后,这两个峰的结合能位置降低到了分别为~69 eV 和~68 eV,可以认为无机溴化物中的溴原子的 XPS 响应[159]。WEEEs 的无机成分分析结果(表 2.8)表明,其中存在大量

的无机元素可作为溴自由基的捕获剂。

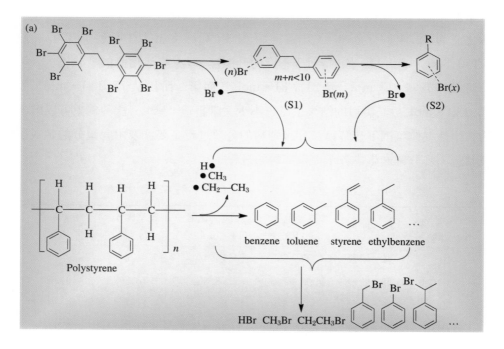

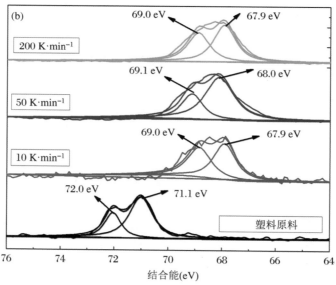

图 2.33 (a) 溴化阻燃剂在热解过程中的分解和转化机理示意图;(b) 不同升温速率条件下 Br XPS 图谱

在本节中,我们对废弃生物质和 WEEEs 塑料共热解过程进行了研究,探索在热解过程中溴化阻燃剂的分解转化过程和机理。根据实验结果,可以得出以下结论:

(1) 共热解过程主要受温度、生物质与 WEEEs 塑料的混合比例以及升温速率的影响。而生物质和 WEEEs 塑料在热解过程中的协同作用可以大大提高热解油的产率。

（2）在共热解过程中，溴元素在热解生物炭、生物油和热解气中都有分布，其中溴元素在热解气体中的分布小于 10%，而其在热解油中的分布随着温度的升高而降低。考虑到溴元素在燃烧过程中会产生大量的有害污染物，因此，生物油将不适合直接作为燃料使用，但是由于热解油中含有大量的芳香烃类化合物，其可以作为化工原料用于提取高附加值化学品。

（3）机理分析表明，热解过程中产生的溴元素既可以被有机组分捕获形成溴代烃类挥发至热解油和气体相中，也可以被 WEEEs 塑料和生物质中的无机组分捕获而保留在热解生物炭中。因此为了尽可能地降低溴元素的挥发，我们可以考虑在热解过程中加入一些无机捕获剂，使溴元素尽可能多地被捕获而保存在热解生物炭中，降低其挥发带来的二次污染。

参考文献

[1] Qiu J. China faces up to groundwater crisis [J]. Nature, 2010, 466 (7304): 308.

[2] Godfray H C J, Beddington J R, Crute I R, et al. Food security: The challenge of feeding 9 billion people [J]. Science, 2010, 327 (5967): 812-818.

[3] Zhang S, Du Q, Cheng K, et al. Efficient phosphorus recycling and heavy metal removal from wastewater sludge by a novel hydrothermal humification-technique [J]. Chemical Engineering Journal, 2020, 394: 124832.

[4] You D, Shi H, Xi Y, et al. Simultaneous heavy metals removal via in situ construction of multivariate metal-organic gels in actual wastewater and the reutilization for Sb(V) capture [J]. Chemical Engineering Journal, 2020, 400: 125359.

[5] Ma D, Su M, Qian J, et al. Heavy metal removal from sewage sludge under citric acid and electroosmotic leaching processes [J]. Separation and Purification Technology, 2020, 242: 116822.

[6] Jawed A, Saxena V, Pandey L M. Engineered nanomaterials and their surface functionalization for the removal of heavy metals: A review [J].

Journal of Water Process Engineering, 2020, 33: 101009.

[7] He M, Wang L, Lv Y, et al. Novel polydopamine/metal organic framework thin film nanocomposite forward osmosis membrane for salt rejection and heavy metal removal [J]. Chemical Engineering Journal, 2020, 389: 124452.

[8] Duan C, Ma T, Wang J, et al. Removal of heavy metals from aqueous solution using carbon-based adsorbents: A review [J]. Journal of Water Process Engineering, 2020, 37: 101339.

[9] Wang J, Chen C. Biosorption of heavy metals by Saccharomyces cerevisiae: A review [J]. Biotechnology Advances, 2006, 24(5): 427-451.

[10] Wang J, Chen C. Biosorbents for heavy metals removal and their future [J]. Biotechnology Advances, 2009, 27(2): 195-226.

[11] Rao R A K, Khan M A, Rehman F. Utilization of fennel biomass (foeniculum vulgari) a medicinal herb for the biosorption of Cd(Ⅱ) from aqueous phase [J]. Chemical Engineering Journal, 2010, 156(1): 106-113.

[12] Özdemir S, Kilinc E, Poli A, et al. Biosorption of Cd, Cu, Ni, Mn and Zn from aqueous solutions by *Thermophilic bacteria*, *Geobacillus toebii* sub. sp. *decanicus* and *Geobacillus thermoleovorans* sub. sp. *stromboliensis*: Equilibrium, kinetic and thermodynamic studies [J]. Chemical Engineering Journal, 2009, 152(1): 195-206.

[13] Oliveira W E, Franca A S, Oliveira L S, et al. Untreated coffee husks as biosorbents for the removal of heavy metals from aqueous solutions [J]. Journal of Hazardous Materials, 2008, 152(3): 1073-1081.

[14] Zhou J, Chen L H, Peng L, et al. Phytoremediation of heavy metals under an oil crop rotation and treatment of biochar from contaminated biomass for safe use [J]. Chemosphere, 2020, 247: 125856.

[15] Yang S S, Chen Y D, Zhang Y, et al. A novel clean production approach to utilize crop waste residues as co-diet for mealworm (*Tenebrio molitor*) biomass production with biochar as byproduct for heavy metal removal [J]. Environmental Pollution, 2019, 252: 1142-1153.

[16] Urrutia C, Yañez-Mansilla E, Jeison D. Bioremoval of heavy metals from metal mine tailings water using microalgae biomass [J]. Algal Research, 2019, 43: 101659.

[17] He J, Strezov V, Zhou X, et al. Pyrolysis of heavy metal contaminated

biomass pre-treated with ferric salts: Product characterisation and heavy metal deportment [J]. Bioresource Technology, 2020, 313: 123641.

[18] Cepoi L, Zinicovscaia I, Rudi L, et al. Growth and heavy metals accumulation by Spirulina platensis biomass from multicomponent copper containing synthetic effluents during repeated cultivation cycles [J]. Ecological Engineering, 2020, 142: 105637.

[19] Gurgel L V A, Gil L F. Adsorption of Cu(II), Cd(II) and Pb(II) from aqueous single metal solutions by succinylated twice-mercerized sugarcane bagasse functionalized with triethylenetetramine [J]. Water Research, 2009, 43(18): 4479-4488.

[20] Das S K, Das A R, Guha A K. A study on the adsorption mechanism of mercury on aspergillus versicolor biomass [J]. Environmental Science & Technology, 2007, 41(24): 8281-8287.

[21] Repo E, Warchol J K, Kurniawan T A, et al. Adsorption of Co(II) and Ni(II) by EDTA- and/or DTPA-modified chitosan: Kinetic and equilibrium modeling [J]. Chemical Engineering Journal, 2010, 161(1/2): 73-82.

[22] Dai J, Yan H, Yang H, et al. Simple method for preparation of chitosan/poly(acrylic acid) blending hydrogel beads and adsorption of copper(II) from aqueous solutions [J]. Chemical Engineering Journal, 2010, 165(1): 240-249.

[23] Luo F, Liu Y, Li X, et al. Biosorption of lead ion by chemically-modified biomass of marine brown algae *Laminaria japonica* [J]. Chemosphere, 2006, 64(7): 1122-1127.

[24] Sillanpää M, Orama M, Rämö J, et al. The importance of ligand speciation in environmental research: A case study [J]. Science of The Total Environment, 2001, 267(1/2/3): 23-31.

[25] Nagib S, Inoue K, Yamaguchi T, et al. Recovery of Ni from a large excess of Al generated from spent hydrodesulfurization catalyst using picolylamine type chelating resin and complexane types of chemically modified chitosan [J]. Hydrometallurgy, 1999, 51(1): 73-85.

[26] Yu J, Tong M, Sun X, et al. Enhanced and selective adsorption of Pb^{2+} and Cu^{2+} by EDTAD-modified biomass of baker's yeast [J]. Bioresource Technology, 2008, 99(7): 2588-93.

[27] Boehm H P. Some aspects of the surface chemistry of carbon blacks and

other carbons [J]. Carbon, 1994, 32(5): 759-769.

[28] Munagapati V S, Yarramuthi V, Nadavala S K, et al. Biosorption of Cu(II), Cd(II) and Pb(II) by *Acacia leucocephala* bark powder: Kinetics, equilibrium and thermodynamics [J]. Chemical Engineering Journal, 2010, 157(2/3): 357-365.

[29] Ho Y S, Mckay G. Pseudo-second order model for sorption processes [J]. Process Biochemistry, 1999, 34(5): 451-465.

[30] Wu F C, Tseng R L, Juang R S. Kinetic modeling of liquid-phase adsorption of reactive dyes and metal ions on chitosan [J]. Water Research, 2001, 35(3): 613-618.

[31] Langmuir I. The adsorption of gases on plane surfaces of glass, mica and platinum [J]. Journal of the American Chemical Society, 1918, 40(9): 1361-1403.

[32] Wang L, Yang L, Li Y, et al. Study on adsorption mechanism of Pb(II) and Cu(II) in aqueous solution using PS-EDTA resin [J]. Chemical Engineering Journal, 2010, 163(3): 364-372.

[33] Freundlich H M F. Uber die adsorption in losungen [J]. Z Phys Chem, 1906, 57A: 385-470.

[34] Plazinski W, Rudzinski W. Modeling the effect of pH on kinetics of heavy metal ion biosorption. A theoretical approach based on the statistical rate theory [J]. Langmuir, 2008, 25(1): 298-304.

[35] Schecher W D, McAvoy D C. MINEQL+: A chemical equilibrium program for personal computers (Version 45) [Z]. Environmental Research Software, Hallowell, Maine, USA, 2001.

[36] Dean J A. Lange's Hand Book of Chemistry [M]. 15th ed. New York: McGraw-Hill Book Company, 1999.

[37] Tan G, Yuan H, Liu Y, et al. Removal of lead from aqueous solution with native and chemically modified corncobs [J]. Journal of Hazardous Materials, 2010, 174(1/2/3): 740-745.

[38] Iqbal M, Saeed A, Zafar S I. FTIR spectrophotometry, kinetics and adsorption isotherms modeling, ion exchange, and EDX analysis for understanding the mechanism of Cd^{2+} and Pb^{2+} removal by mango peel waste [J]. Journal of Hazardous Materials, 2009, 164(1): 161-171.

[39] Zheng J C, Feng H M, Lam M H W, et al. Removal of Cu(II) in aqueous

[40] Yang F, Liu H J, Qu J H, et al. Preparation and characterization of chitosan encapsulated *Sargassum* sp. biosorbent for nickel ions sorption [J]. Bioresource Technology, 2011, 102(3): 2821-2828.

[41] Lim S F, Zheng Y M, Zou S W, et al. Characterization of copper adsorption onto an alginate encapsulated magnetic sorbent by a combined FT-IR, XPS, and mathematical modeling study [J]. Environmental Science & Technology, 2008, 42(7): 2551-2556.

[42] Toupin M, Belanger D. Spontaneous functionalization of carbon black by reaction with 4-nitrophenyldiazonium cations [J]. Langmuir, 2008, 24(5): 1910-1917.

[43] Zhou J H, Sui Z J, Zhu J, et al. Characterization of surface oxygen complexes on carbon nanofibers by TPD, XPS and FT-IR [J]. Carbon, 2007, 45(4): 785-796.

[44] Vázquez G, Calvo M, Sonia Freire M, et al. Chestnut shell as heavy metal adsorbent: Optimization study of lead, copper and zinc cations removal [J]. Journal of Hazardous Materials, 2009, 172(2/3): 1402-1414.

[45] Godelitsas A, Astilleros J M, Hallam K, et al. Interaction of calcium carbonates with lead in aqueous solutions [J]. Environmental Science & Technology, 2003, 37(15): 3351-3360.

[46] Sud D, Mahajan G, Kaur M. Agricultural waste material as potential adsorbent for sequestering heavy metal ions from aqueous solutions: A review [J]. Bioresource Technology, 2008, 99(14): 6017-6027.

[47] Farooq U, Kozinski J A, Khan M A, et al. Biosorption of heavy metal ions using wheat based biosorbents: A review of the recent literature [J]. Bioresource Technology, 2010, 101(14): 5043-5053.

[48] Sunsandee N, Ramakul P, Phatanasri S, et al. Biosorption of dicloxacillin from pharmaceutical waste water using tannin from Indian almond leaf: Kinetic and equilibrium studies [J]. Biotechnology Reports, 2020, 27: e00488.

[49] Rocha De Freitas G, Adeodato Vieira M G, Carlos Da Silva M G. Characterization and biosorption of silver by biomass waste from the alginate industry [J]. Journal of Cleaner Production, 2020, 271: 122588.

[50] Rambabu K, Bharath G, Banat F, et al. Biosorption performance of date palm empty fruit bunch wastes for toxic hexavalent chromium removal [J]. Environmental Research, 2020, 187: 109694.

[51] Jin Z, Deng S, Wen Y, et al. Application of Simplicillium chinense for Cd and Pb biosorption and enhancing heavy metal phytoremediation of soils [J]. Science of The Total Environment, 2019, 697: 134148.

[52] Fathollahi A, Coupe S J, El-Sheikh A H, et al. The biosorption of mercury by permeable pavement biofilms in stormwater attenuation [J]. Science of the Total Environment, 2020, 741: 140411.

[53] Beni A A, Esmaeili A. Biosorption, an efficient method for removing heavy metals from industrial effluents: A review [J]. Environmental Technology & Innovation, 2020, 17: 100503.

[54] Jin Z, Xie L, Zhang T, et al. Interrogating cadmium and lead biosorption mechanisms by Simplicillium chinense via infrared spectroscopy [J]. Environmental Pollution, 2020, 263: 114419.

[55] Park D, Yun Y-S, Jo J H, et al. Biosorption process for treatment of electroplating wastewater containing Cr (Ⅳ): Laboratory-scale feasibility test [J]. Industrial & Engineering Chemistry Research, 2006, 45(14): 5059-5065.

[56] Vilar V J P, Botelho C M S, Boaventura R a R. Lead and copper biosorption by marine red algae gelidium and algal composite material in a CSTR ("Carberry" type) [J]. Chemical Engineering Journal, 2008, 138(1/2/3): 249-257.

[57] Mata Y N, Blázquez M L, Ballester A, et al. Sugar-beet pulp pectin gels as biosorbent for heavy metals: Preparation and determination of biosorption and desorption characteristics [J]. Chemical Engineering Journal, 2009, 150(2/3): 289-301.

[58] Koppolu L, Agblevor F A, Clements L D. Pyrolysis as a technique for separating heavy metals from hyperaccumulators. Part Ⅱ: Lab-scale pyrolysis of synthetic hyperaccumulator biomass [J]. Biomass and Bioenergy, 2003, 25(6): 651-663.

[59] Koppolu L, Prasad R, Davis Clements L. Pyrolysis as a technique for separating heavy metals from hyperaccumulators. Part Ⅲ: Pilot-scale pyrolysis of synthetic hyperaccumulator biomass [J]. Biomass and

Bioenergy, 2004, 26(5): 463-472.

[60] Stals M, Carleer R, Reggers G, et al. Flash pyrolysis of heavy metal contaminated hardwoods from phytoremediation: Characterisation of biomass, pyrolysis oil and char/ash fraction [J]. Journal of Analytical and Applied Pyrolysis, 2010, 89(1): 22-29.

[61] Stals M, Thijssen E, Vangronsveld J, et al. Flash pyrolysis of heavy metal contaminated biomass from phytoremediation: Influence of temperature, entrained flow and wood/leaves blended pyrolysis on the behaviour of heavy metals [J]. Journal of Analytical and Applied Pyrolysis, 2010, 87(1): 1-7.

[62] Zeng K, Li R, Minh D P, et al. Characterization of char generated from solar pyrolysis of heavy metal contaminated biomass [J]. Energy, 2020, 206: 118128.

[63] Zeng K, Li R, Minh D P, et al. Solar pyrolysis of heavy metal contaminated biomass for gas fuel production [J]. Energy, 2019, 187: 116016.

[64] He J, Strezov V, Kumar R, et al. Pyrolysis of heavy metal contaminated avicennia marina biomass from phytoremediation: Characterisation of biomass and pyrolysis products [J]. Journal of Cleaner Production, 2019, 234: 1235-1245.

[65] He J, Strezov V, Kan T, et al. Effect of temperature on heavy metal(loid) deportment during pyrolysis of avicennia marina biomass obtained from phytoremediation [J]. Bioresource Technology, 2019, 278: 214-222.

[66] Han Z, Guo Z, Zhang Y, et al. Adsorption-pyrolysis technology for recovering heavy metals in solution using contaminated biomass phytoremediation [J]. Resources, Conservation and Recycling, 2018, 129: 20-26.

[67] Westerh of R J M, Brilman D W F, Van Swaaij W P M, et al. Effect of temperature in fluidized bed fast pyrolysis of biomass: Oil quality assessment in test units [J]. Industrial & Engineering Chemistry Research, 2009, 49(3): 1160-1168.

[68] Mehrabani J V, Noaparast M, Mousavi S M, et al. Depression of pyrite in the flotation of high pyrite low-grade lead-zinc ore using Acidithiobacillus ferrooxidans [J]. Minerals Engineering, 2010, 23(1): 10-16.

[69] Schindler D W, Hecky R E. Eutrophication: More nitrogen data needed [J]. Science, 2009, 324(5928): 721-722.

[70] Zhu G, Peng Y, Zhai L, et al. Performance and optimization of biological nitrogen removal process enhanced by anoxic/oxic step feeding [J]. Biochemical Engineering Journal, 2009, 43(3): 280-287.

[71] Kumar M, Lin J G. Co-existence of anammox and denitrification for simultaneous nitrogen and carbon removal: strategies and issues [J]. Journal of Hazardous Materials, 2010, 178(1/2/3): 1-9.

[72] Yang S, Yang F, Fu Z, et al. Simultaneous nitrogen and phosphorus removal by a novel sequencing batch moving bed membrane bioreactor for wastewater treatment [J]. Journal of Hazardous Materials, 2010, 175(1/2/3): 551-557.

[73] Guo C H, Stabnikov V, Ivanov V. The removal of nitrogen and phosphorus from reject water of municipal wastewater treatment plant using ferric and nitrate bioreductions [J]. Bioresource Technology, 2010, 101(11): 3992-3999.

[74] Vohla C, Alas R, Nurk K, et al. Dynamics of phosphorus, nitrogen and carbon removal in a horizontal subsurface flow constructed wetland [J]. Science of the Total Environment, 2007, 380(1-3): 66-74.

[75] Yousefi Z, Mohseni-Bandpei A. Nitrogen and phosphorus removal from wastewater by subsurface wetlands planted with *Iris pseudacorus* [J]. Ecological Engineering, 2010, 36(6): 777-782.

[76] Zhao C, Liu S, Jiang Z, et al. Nitrogen purification potential limited by nitrite reduction process in coastal eutrophic wetlands [J]. Science of the Total Environment, 2019, 694: 133702.

[77] Xing W, Han Y, Guo Z, et al. Quantitative study on redistribution of nitrogen and phosphorus by wetland plants under different water quality conditions [J]. Environmental Pollution, 2020, 261: 114086.

[78] Vymazal J, Sochacki A, Fučík P, et al. Constructed wetlands with subsurface flow for nitrogen removal from tile drainage [J]. Ecological Engineering, 2020, 155: 105943.

[79] Steidl J, Kalettka T, Bauwe A. Nitrogen retention efficiency of a surface-flow constructed wetland receiving tile drainage water: A case study from north-eastern Germany [J]. Agriculture, Ecosystems & Environment, 2019, 283: 106577.

[80] Rampuria A, Gupta A B, Brighu U. Nitrogen transformation processes and

mass balance in deep constructed wetlands treating sewage, exploring the anammox contribution [J]. Bioresource Technology, 2020, 314: 123737.

[81] Nilsson J E, Liess A, Ehde P M, et al. Mature wetland ecosystems remove nitrogen equally well regardless of initial planting [J]. Science of The Total Environment, 2020, 716: 137002.

[82] Liu C, Hou L, Liu M, et al. In situ nitrogen removal processes in intertidal wetlands of the Yangtze Estuary [J]. Journal of Environmental Sciences, 2020, 93: 91-97.

[83] Jia J, Bai J, Gao H, et al. Different effects of NaCl and Na_2SO_4 on soil net nitrogen mineralization in coastal wetlands [J]. Ecotoxicology and Environmental Safety, 2020, 199: 110678.

[84] Spieles D J, Mitsch W J. The effects of season and hydrologic and chemical loading on nitrate retention in constructed wetlands: A comparison of low- and high-nutrient riverine systems [J]. Ecological Engineering, 1999, 14(1/2): 77-91.

[85] Hart M R, Quin B F, Nguyen M L. Phosphorus runoff from agricultural land and direct fertilizer effects [J]. Journal of Environment Quality, 2004, 33(6): 1954-1972.

[86] Longhi D, Bartoli M, Viaroli P. Decomposition of four macrophytes in wetland sediments: Organic matter and nutrient decay and associated benthic processes [J]. Aquatic Botany, 2008, 89(3): 303-310.

[87] Kuehn K A, Suberkropp K. Decomposition of standing litter of the freshwater emergent macrophyteJuncus effusus [J]. Freshwater Biology, 1998, 40(4): 717-727.

[88] Masi F. Water reuse and resources recovery: The role of constructed wetlands in the Ecosan approach [J]. Desalination, 2009, 246(1/2/3): 27-34.

[89] Meuleman A F M, Van Logtestijn R, Rijs G B J, et al. Water and mass budgets of a vertical-flow constructed wetland used for wastewater treatment [J]. Ecological Engineering, 2003, 20(1): 31-44.

[90] Verma V K, Singh Y P, Rai J P N. Biogas production from plant biomass used for phytoremediation of industrial wastes [J]. Bioresource Technology, 2007, 98(8): 1664-1669.

[91] Leterme P, Londoño A M, Muñoz J E, et al. Nutritional value of aquatic ferns (azolla filiculoides lam. and salvinia molesta mitchell) in pigs [J]. Animal Feed Science and Technology, 2009, 149(1/2): 135-148.

[92] Kalita P, Mukhopadhyay P K, Mukherjee A K. Evaluation of the nutritional quality of four unexplored aquatic weeds from northeast India for the formulation of cost-effective fish feeds [J]. Food Chemistry, 2007, 103(1): 204-209.

[93] Czernik S, Bridgwater A V. Overview of applications of biomass fast pyrolysis oil [J]. Energy & Fuels, 2004, 18(2): 590-598.

[94] Zheng S, Yang Z, Sun M. Pollutant removal from municipal sewage in winter via a modified free-water-surface system planted with edible vegetable [J]. Desalination, 2010, 250(1): 158-161.

[95] Dupont C, Commandré J-M, Gauthier P, et al. Biomass pyrolysis experiments in an analytical entrained flow reactor between 1073 K and 1273 K [J]. Fuel, 2008, 87(7): 1155-1164.

[96] Friedl A, Padouvas E, Rotter H, et al. Prediction of heating values of biomass fuel from elemental composition [J]. Analytica Chimica Acta, 2005, 544(1/2): 191-198.

[97] Desisto W J, Hill N, Beis S H, et al. Fast pyrolysis of pine sawdust in a fluidized-bed reactor [J]. Energy & Fuels, 2010, 24(4): 2642-2651.

[98] Wang P, Zhan S, Yu H, et al. The effects of temperature and catalysts on the pyrolysis of industrial wastes (herb residue) [J]. Bioresource Technology, 2010, 101(9): 3236-3241.

[99] Heo H S, Park H J, Park Y-K, et al. Bio-oil production from fast pyrolysis of waste furniture sawdust in a fluidized bed [J]. Bioresource Technology, 2010, 101(1): S91-S96.

[100] Boateng A A, Cooke P H, Hicks K B. Microstructure development of chars derived from high-temperature pyrolysis of barley (*Hordeum vulgare* L.) hulls [J]. Fuel, 2007, 86(5/6): 735-742.

[101] Ren Q, Zhao C. NO_x and N_2O Precursors from biomass pyrolysis: Nitrogen transformation from amino acid [J]. Environmental Science & Technology, 2012, 46(7): 4236-4240.

[102] Becidan M, Skreiberg O, Hustad J E. NO_x and N_2O precursors (NH_3 and HCN) in pyrolysis of biomass residues [J]. Energy & Fuels, 2007, 21(2):

1173-1180.

[103] Al-Salem S M, Lettieri P, Baeyens J. Recycling and recovery routes of plastic solid waste (PSW): A review [J]. Waste Management, 2009, 29 (10): 2625-2643.

[104] Zhao Y B, Lv X D, Ni H G. Solvent-based separation and recycling of waste plastics: A review [J]. Chemosphere, 2018, 209: 707-720.

[105] Zhang J-P, Zhang F-S. A new approach for blending waste plastics processing: Superabsorbent resin synthesis [J]. Journal of Cleaner Production, 2018, 197: 501-510.

[106] Singh N, Duan H, Tang Y. Toxicity evaluation of E-waste plastics and potential repercussions for human health [J]. Environment International, 2020, 137: 105559.

[107] Needhidasan S, Vigneshwar C R, Ramesh B. Amalgamation of E-waste plastics in concrete with super plasticizer for better strength [J]. Materials Today: Proceedings, 2020, 22: 998-1003.

[108] Needhidasan S, Ramesh B, Joshua Richard Prabu S. Experimental study on use of E-waste plastics as coarse aggregate in concrete with manufactured sand [J]. Materials Today: Proceedings, 2020, 22: 715-721.

[109] Mir R A, Singla S, Pandey O P. Hetero carbon structures derived from waste plastics as an efficient electrocatalyst for water splitting and high-performance capacitors [J]. Physica E: Low-dimensional Systems and Nanostructures, 2020, 124: 114284.

[110] Li Y, Chang Q, Luo Z, et al. Transfer of POP-BFRs within E-waste plastics in recycling streams in China [J]. Science of The Total Environment, 2020, 717: 135003.

[111] Chen Y, Cui Z, Cui X, et al. Life cycle assessment of end-of-life treatments of waste plastics in China [J]. Resources, Conservation and Recycling, 2019, 146: 348-357.

[112] Robinson B H. E-waste: An assessment of global production and environmental impacts [J]. Science of The Total Environment, 2009, 408 (2): 183-191.

[11] Tukker A, De Groot H, Simons L, et al. Chemical recycling of plastic waste (PVC and other resins) [C]. European Commission, DG III, Final Report, STB-99-55 Final Del, The Netherlands, 1999.

[114] Taurino R, Pozzi P, Zanasi T. Facile characterization of polymer fractions from Waste Electrical and Electronic Equipment (WEEE) for mechanical recycling [J]. Waste Management, 2010, 30(12): 2601-2607.

[115] Choi J, Kim O, Kwak S-Y. Suppression of dioxin emission in co-incineration of poly(vinyl chloride) with TiO_2-encapsulating polystyrene [J]. Environmental Science & Technology, 2007, 41(16): 5833-5838.

[116] Wäger P A, Schluep M, Müller E, et al. RoHS regulated substances in mixed plastics from waste electrical and electronic equipment [J]. Environmental Science & Technology, 2011, 46(2): 628-635.

[117] Tian H, Gao J, Lu L, et al. Temporal trends and spatial variation characteristics of hazardous air pollutant emission inventory from municipal solid waste incineration in China [J]. Environmental Science & Technology, 2012, 46(18): 10364-10371.

[118] Baytekin B, Baytekin H T, Grzybowski B A. Retrieving and converting energy from polymers: Deployable technologies and emerging concepts [J]. Energy & Environmental Science, 2013, 6: 3467-3482.

[119] Alston S M, Arnold J C. Environmental impact of pyrolysis of mixed WEEE plastics. Part 2: Life cycle assessment [J]. Environmental Science & Technology, 2011, 45(21): 9386-9392.

[120] Zhang Y, Duan D, Lei H, et al. Jet fuel production from waste plastics via catalytic pyrolysis with activated carbons [J]. Applied Energy, 2019, 251: 113337.

[121] Xu S, Cao B, Uzoejinwa B B, et al. Synergistic effects of catalytic co-pyrolysis of macroalgae with waste plastics [J]. Process Safety and Environmental Protection, 2020, 137: 34-48.

[122] Kasar P, Sharma D K, Ahmaruzzaman M. Thermal and catalytic decomposition of waste plastics and its co-processing with petroleum residue through pyrolysis process [J]. Journal of Cleaner Production, 2020, 265: 121639.

[123] Jin Z, Yin L, Chen D, et al. Co-pyrolysis characteristics of typical components of waste plastics in a falling film pyrolysis reactor [J]. Chinese Journal of Chemical Engineering, 2018, 26(10): 2176-2184.

[124] Huo E, Lei H, Liu C, et al. Jet fuel and hydrogen produced from waste plastics catalytic pyrolysis with activated carbon and MgO [J]. Science of

the Total Environment, 2020, 727: 138411.

[125] Cai N, Yang H, Zhang X, et al. Bimetallic carbon nanotube encapsulated Fe-Ni catalysts from fast pyrolysis of waste plastics and their oxygen reduction properties [J]. Waste Management, 2020, 109: 119-126.

[126] Bhaskar T, Matsui T, Kaneko J, et al. Novel calcium based sorbent (Ca-C) for the dehalogenation (Br, Cl) process during halogenated mixed plastic (PP/PE/PS/PVC and HIPS-Br) pyrolysis [J]. Green Chemistry, 2002, 4(4): 372-375.

[127] Yang X, Sun L, Xiang J, et al. Pyrolysis and dehalogenation of plastics from Waste Electrical and Electronic Equipment (WEEE): A review [J]. Waste Management, 2013, 33(2): 462-473.

[128] Alston S M, Clark A D, Arnold J C, et al. Environmental impact of pyrolysis of mixed WEEE plastics. Part 1: Experimental pyrolysis data [J]. Environmental Science & Technology, 2011, 45(21): 9380-9385.

[129] Undri A, Frediani M, Rosi L, et al. Reverse polymerization of waste polystyrene through microwave assisted pyrolysis [J]. Journal of Analytical and Applied Pyrolysis, 2014, 105: 35-42.

[130] Wu F, Guo J, Chang H, et al. Polybrominated diphenyl ethers and decabromodiphenylethane in sediments from twelve lakes in China [J]. Environmental Pollution, 2012, 162: 262-268.

[131] Abdullah H, Mediaswanti K A, Wu H. Biochar as a fuel. 2. Significant differences in fuel quality and ash properties of biochars from various biomass components of mallee trees [J]. Energy & Fuels, 2010, 24(3): 1972-1979.

[132] Fu P, Hu S, Xiang J, et al. Pyrolysis of maize stalk on the characterization of chars formed under different devolatilization conditions [J]. Energy & Fuels, 2009, 23(9): 4605-4611.

[133] Salehi E, Abedi J, Harding T. Bio-oil from sawdust: Pyrolysis of sawdust in a fixed-bed system [J]. Energy & Fuels, 2009, 23(7): 3767-3772.

[134] Sonobe T, Worasuwannarak N, Pipatmanomai S. Synergies in co-pyrolysis of Thai lignite and corncob [J]. Fuel Processing Technology, 2008, 89(12): 1371-1378.

[135] Li C Z, Nelson P F. Fate of aromatic ring systems during thermal cracking of tars in a fluidized-bed reactor [J]. Energy & Fuels, 1996, 10(5):

1083-1090.

[136] Liu W J, Zeng F X, Jiang H, et al. Total recovery of nitrogen and phosphorus from three wetland plants by fast pyrolysis technology [J]. Bioresource Technology, 2011, 102(3): 3471-3479.

[137] Yuan S, Chen X L, Li W F, et al. Nitrogen conversion under rapid pyrolysis of two types of aquatic biomass and corresponding blends with coal [J]. Bioresource Technology, 2011, 102(21): 10124-10130.

[138] Brebu M, Jakab E, Sakata Y. Effect of flame retardants and Sb_2O_3 synergist on the thermal decomposition of high-impact polystyrene and on its debromination by ammonia treatment [J]. Journal of Analytical and Applied Pyrolysis, 2007, 79(1/2): 346-352.

[139] Hall W J, Mitan N M M, Bhaskar T, et al. The co-pyrolysis of flame retarded high impact polystyrene and polyolefins [J]. Journal of Analytical and Applied Pyrolysis, 2007, 80(2): 406-415.

[140] Jung S H, Kim S J, Kim J S. Fast pyrolysis of a waste fraction of High Impact Polystyrene (HIPS) containing brominated flame retardants in a fluidized bed reactor: The effects of various Ca-based additives (CaO, $Ca(OH)_2$ and oyster shells) on the removal of bromine [J]. Fuel, 2012, 95: 514-520.

[141] Jakab E, Uddin M A, Bhaskar T, et al. Thermal decomposition of flame-retarded high-impact polystyrene [J]. Journal of Analytical and Applied Pyrolysis, 2003, 68-69: 83-99.

[142] Peil S, Seisel S, Schrems O. FTIR-spectroscopic studies of polar stratospheric cloud model surfaces: Characterization of nitric acid hydrates and heterogeneous reactions involving N_2O_5 and HBr [J]. Journal of Molecular Structure, 1995, 348: 449-452.

[143] Gomez L, Tran H, Jacquemart D. Line mixing calculation in the $\nu 6$ Q-branches of N_2-broadened CH_3Br at low temperatures [J]. Journal of Molecular Spectroscopy, 2009, 256(1): 35-40.

[144] Edreis E M A, Luo G, Li A, et al. CO_2 co-gasification of lower sulphur petroleum coke and sugar cane bagasse via TG-FTIR analysis technique [J]. Bioresource Technology, 2013, 136: 595-603.

[145] Wang Z, Guo Q, Liu X, et al. Low temperature pyrolysis characteristics of oil sludge under various heating conditions [J]. Energy & Fuels, 2007, 21

(2): 957-962.

[146] Chanunpanich N, Ulman A, Strzhemechny Y M, et al. Surface modification of polyethylene through bromination [J]. Langmuir, 1999, 15(6): 2089-2094.

[147] Majumder P, Paul P, Sengupta P, et al. Formation of organopalladium complexes via C—Br and C—C bond activation. Application in C—C and C—N coupling reactions [J]. Journal of Organometallic Chemistry, 2013, 736: 1-8.

[148] Undri A, Frediani M, Rosi L, et al. Reverse polymerization of waste polystyrene through microwave assisted pyrolysis [J]. Journal of Analytical and Applied Pyrolysis, 2014, 105: 35-42.

[149] Kannan P, Biernacki J J, Visco Jr D P, et al. Kinetics of thermal decomposition of expandable polystyrene in different gaseous environments [J]. Journal of Analytical and Applied Pyrolysis, 2009, 84(2): 139-144.

[150] Birch A M, Groombridge S, Law R, et al. Rationally designing safer anilines: The challenging case of 4-aminobiphenyls [J]. Journal of Medicinal Chemistry, 2012, 55(8): 3923-3933.

[151] Vishnoi S, Agrawal V, Kasana V K. Synthesis and structure-activity relationships of substituted cinnamic acids and amide analogues: A new class of herbicides [J]. Journal of Agricultural and Food Chemistry, 2009, 57(8): 3261-3265.

[152] Yazdanbakhsh M R, Mohammadi A, Mohajerani E, et al. Novel azo disperse dyes derived from N-benzyl-N-ethyl-aniline: Synthesis, solvatochromic and optical properties [J]. Journal of Molecular Liquids, 2010, 151(2/3): 107-112.

[153] Ahamad T, Alshehri S M. TG-FTIR-MS (Evolved Gas Analysis) of bidi tobacco powder during combustion and pyrolysis [J]. Journal of Hazardous Materials, 2012, 199/200: 200-208.

[54] Feng Y, Jiang X, Chi Y, et al. Volatilization behavior of fluorine in fluoroborate residue during pyrolysis [J]. Environmental Science & Technology, 2011, 46(1): 307-311.

[155] Brown M E, Maciejewski M, Vyazovkin S, et al. Computational aspects of kinetic analysis. Part A: The ICTAC kinetics project-data, methods and results [J]. Thermochimica Acta, 2000, 355(1/2): 125-143.

[156] Jung S H, Kim S J, Kim J S. The influence of reaction parameters on characteristics of pyrolysis oils from waste high impact polystyrene and acrylonitrile-butadiene-styrene using a fluidized bed reactor [J]. Fuel Processing Technology, 2013, 116: 123-129.

[157] Grause G, Karakita D, Ishibashi J, et al. TG-MS investigation of brominated products from the degradation of brominated flame retardants in high-impact polystyrene [J]. Chemosphere, 2011, 85(3): 368-373.

[158] Hutson N D, Attwood B C, Scheckel K G. XAS and XPS characterization of mercury binding on brominated activated carbon [J]. Environmental Science & Technology, 2007, 41(5): 1747-1752.

[159] Zhang Y C, Tang J Y, Zhou W D. Green hydrothermal synthesis and optical properties of cuprous bromide nanocrystals [J]. Materials Chemistry and Physics, 2008, 108(1): 4-7.

第 3 章

生物质热解液体产物的品质提升及其机理研究

3.1 电催化 5-羟甲基糠醛的阳极氧化制备 2,5-呋喃二甲酸

3.1.1 概述

5-羟甲基糠醛(HMF)是研究较广泛的生物质衍生平台分子之一[1]。HMF 可选择性地氧化成 2,5-二甲酰基呋喃(DFF)[2]、5-羟甲基-2-呋喃羧酸(HFCA)[3]、顺丁烯二酸酐(MA)[4]以及 2,5-呋喃二羧酸(FDCA)[5-10]。FDCA 作为合成生物基聚合物(包括聚乙烯 2,5-呋喃二甲酸酯(PEF))的单体越来越受到关注。[11-14]PEF 是一种可再生聚合物,目前被用作石油衍生的聚对苯二甲酸乙二醇酯塑料(PET)的替代品。[15-16]PEF 与 PET 有相似的结构,其中 FDCA 取代对苯二甲酸。与 PET 相比,PEF 改善了渗透性能,CO_2 和 O_2 渗透率分别降低了 19 倍和 10 倍,其力学和热性能与 PET 相似。HMF 通常在高压空气或 O_2(如 0.3~2.0 MPa)、高温(30~130 ℃)下的碱性水溶液(pH≥13)中被氧化成 FDCA,使用贵金属基催化剂(如 Au、Pt、Ru 和 Pd)。[17-19]例如,Wan 等人报道,碳纳米管(CNT)催化剂上负载的金-钯催化剂可以在 100 ℃、0.5 MPa O_2 条件下选择性地将 HMF(25 mmol·L^{-1})氧化成 FDCA。[20] Yi 等人研究了在商业 Ru/C 催化剂上,在 120 ℃、0.2 MPa O_2 下催化氧化 HMF(100 mmol·L^{-1}),10 h 后得到 85% 的 FDCA。[21]相对而言,人们更希望研究出改进的方法,使用条件更温和与价格更便宜的催化剂来生产 FDCA。

电化学 HMF 氧化是一种很有前途的替代传统的多相催化好氧氧化的方法,因为它通常是在环境温度下进行的,由阳极电极上的电位驱动,从而避免使用 O_2 或其他危险的化学氧化剂。[22]电化学氧化可以更方便地在较小规模的分散反应器中进行。随着可再生能源(如太阳能和风能)的发电成本持续下降,它的成本竞争力也将越来越强。在电化学羟甲基糠醛氧化过程中,表面反应可以通过应用电位来调节。但是目前电化学 HMF 氧化的有效催化剂仅限于贵金属

基催化剂(Pt,Au,Ru 和 Pd)[9-10,22],选择性较低。例如,Chadderdon 等人研究了碳载金、钯纳米颗粒在碱性介质中对羟甲基糠醛的电催化氧化。结果表明,HMF(20 mmol·L^{-1})在 0.9 V(vs RHE)电位下可在 1 h 内被完全氧化,但 FDCA 选择性最高仅为 83%。[9]在 Cha 和 Choi 的一项研究中[22],以 2,2,6,6-四甲基哌啶-1-氧基(TEMPO)为介质,在 1.54 V(vs RHE)的 Au 电极上,HMF 可以以 100%的法拉第效率定量地转化为 FDCA。然而,使用 TEMPO 会增加下游分离成本。

近年来,人们开发了地球上存在丰富的过渡金属磷化物和硫化物用于羟甲基糠醛的电化学氧化的方法。例如,Sun 及其同事使用 Ni_2P,Ni_2S_3 甚至金属镍作为 HMF 氧化的阳极催化剂,结果 HMF 以 98%至 100%的法拉第效率几乎定量地转化为 FDCA。[7-8,23-24]然而,金属硫化物和磷化物在氧化电位下的热力学稳定性通常不如相应的氧化物/氢氧化物,而在电催化条件下可能会形成金属氧化物/氢氧化物,这意味着活性物质实际上可能不是磷化物和硫化物,特别是在强氧化环境的水溶液中,如水氧化。[25-27]与 Ir 和 Ru 基氧化物相比,含 Ni,Co 和 Fe 的过渡金属层状双氢氧化物(LDH)或一般的金属氢氧化物作为碱性电解质的水氧化催化剂,其起始过电位小于 300 mV,且具有良好的稳定性。[28-29] LDHs 是天然水滑石矿物的合成类似物,具有类似水镁石的带正电荷的混合金属氢氧化物层,层间插有水和电荷平衡的阴离子。[30]它们的合成过程简单,可以直接生长在三维导电衬底上,如碳纤维纸和泡沫镍,[31-32]并且层的剥离进一步提高了水的氧化性能。[31,33]在各种二元 LDH 中,NiFe 材料被认为是最有前途的水氧化电催化剂,尽管最近三元组成显示出更低的过电位以达到类似的电流密度。[32,34-35]然而,有了这些催化剂,这一具有挑战性的四电子质子耦合电子转移反应仍存在水氧化的动力学障碍,活性催化位点的识别仍存在争议。[36-37]

在这项工作中,我们研究并证明了 NiFe LDH 纳米片是高效的阳极电催化剂,用于氧化 HMF 到 FDCA。采用水热法在碳纤维纸上生长 NiFe LDH 纳米片。以羟甲基糠醛为原料,在 1.23 V(vs RHE)电位下直接氧化制备 FDCA,产率为 98%,法拉第效率为 99.4%。此外,由于 HMF 氧化在动力学上比水氧化更有利,因此它可以作为一种替代阳极反应,在水裂解电池中促进 H_2 的产生,并氧化产生高价值有机物。[38]

3.1.2
阳极电催化剂的合成及结构特征

采用水热法直接在经亲水性处理后的碳纤维纸上合成了 NiFe LDH 纳米片。如图 3.1 所示,NiFe LDH 纳米片的生长主要包括两个步骤:① Ni^{2+} 和 Fe^{2+} 与弱碱三乙醇胺(TEOA)的络合,通过络合也可以帮助保护 Fe^{2+},避免其过度氧化至 Fe^{3+} 和 ② 金属氢氧化物的成核和增长,由于 NH_3 的缓慢释放,控制水解,从而在低过饱和条件下控制晶体生长,促进了超薄 LDH 纳米片的形成。[31,36] 通过扫描电子显微镜分析了电极材料的形貌。如图 3.1(a)所示,所制备的电极材料由纳米片组成,纳米片垂直排列在导电碳纤维上,并完全覆盖在基体上。观察到一些单独的六角形片,符合 LDH 晶体的基本特征。[39-40] EDX 元素分析证实了 Ni,Fe,O 均匀分布在碳纤维纸上。NiFe LDH 的 XRD 谱图显示了与天然 NiFe LDH 矿物 Reevesite ICSD ♯107625 相匹配的衍射峰。[41] Ni 2p XPS 峰呈现出两个自旋轨道峰(识别为 Ni $2p_{3/2}$ 和 $2p_{1/2}$)以及两个震荡卫星峰。在结合能为 855.48 eV 及 873.07 eV 的峰可以被归因为 NiO 中的 Ni(Ⅱ),而在结合能为 857.79 eV 和 874.95 eV 的峰可以被归因为 $Ni(OH)_2$ 中的 Ni(Ⅱ)。[28,42] 至于 Fe_{2p} XPS 谱,在结合能为 712.20 eV 及 725.00 eV 的两个大峰可以被归因为 Fe(Ⅱ)。而在结合能为 715.12 eV 及 727.79 eV 的两个小峰可以被归因为 Fe(Ⅲ)。[43]

3.1.3
5-羟甲基糠醛的电化学氧化

我们将 NiFe LDH 纳米片直接作为改性 H 型电化学池的阳极电极催化剂,在导电碳纸上进行 5-羟甲基糠醛的电催化氧化性能评价,如图 3.2 所示。氧化的第一步是将 HMF 转化为 2,5-二甲酰呋喃(DFF)或 5-羟甲基-2-呋喃甲酸(HMFCA)。DFF 和 HMFCA 均可进一步氧化成 5-甲酰基-2-呋喃甲酸(FFCA),再氧化成 FDCA。[22] 水氧化是主要的竞争反应。[23]

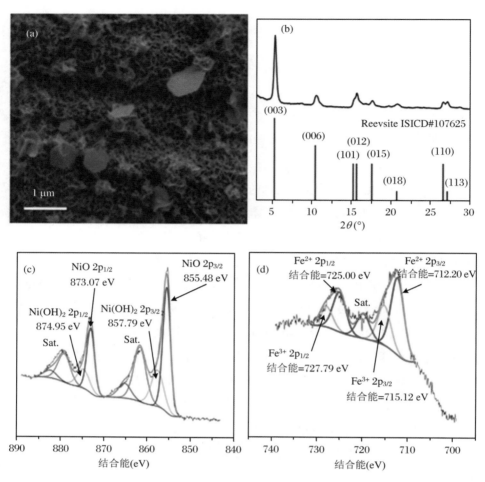

图 3.1 (a) NiFe LDH 纳米片材料的 SEM 图;(b) XRD 图;(c) XPS Ni 2p 谱图;(d) XPS Fe 2p 谱图

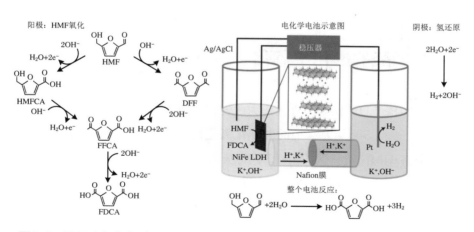

图 3.2 HMF 电极氧化过程示意图
包括阳极反应、阴极反应以及整个电池反应

图 3.3 对比了 NiFe LDH 电极和原始碳纤维纸在 1 mmol·L^{-1} KOH 溶液中 HMF 氧化(10 mmol·L^{-1})和水氧化(无 HMF)的线性扫描伏安曲线。NiFe LDH 在无 HMF 电解液中表现出 1.37 V (vs RHE)的水氧化电位,并在 1.53 V 电位下达到 20 mA·cm^{-2} 的电流密度,与同类 LDH 电极的性能相当。[36] 相比较而言,在 1.25 V 时,HMF 氧化的起始电位较低,在 1.32 V 时电流密度可达到 20 mA·cm^{-2}。这说明在较低的作用电位下,HMF 氧化比水氧化更容易发生。原始碳纤维对 HMF 和水氧化的活性都很低(表 3.1)。水和 HMF 氧化的 Tafel 分析如图 3.3(b)所示。HMF 氧化的 Tafel 斜率为 75 mV·dec^{-1},远低于水氧化的 Tafel 斜率(143 mV·dec^{-1}),进一步证实了在这些电位下,HMF 氧化的速度要快于水氧化。需要指出的是,本研究中水氧化的 Tafel 斜率 (143 mV·dec^{-1})高于一般报道的值(一般为 30~65 mV,在三电极电池中测定)。[44-45] Tafel 斜率高的原因是用于分离电池正极和负极的 Nafion 膜(见图 3.2)在电化学过程中含有相当大的膜电阻,导致出现较大的表观 Tafel 斜率。

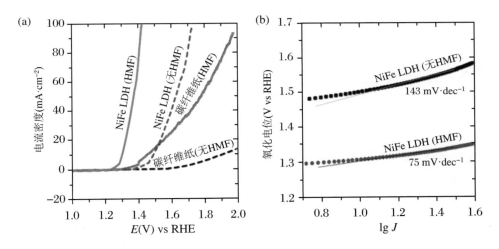

图 3.3 (a) LSV 曲线;(b) Tafel 斜率

采用 HPLC 法监测电化学反应过程中反应物浓度和产物浓度,结果如图 3.4(a)所示。除 FDCA 和 HMF 外,还观察到一个小峰为 FFCA。而传统的多相催化好氧 HMF 氧化和贵金属催化电化学 HMF 氧化的主要副产物是 DFF 和 HMFCA 并未被检测到。[7,22,46] 随着计时电流测试的进行,HMF 浓度下降,阳极电解液的颜色由藏红花黄色变为无色(图 3.4(b))。HMF 及其氧化产物的浓度随计时电流测试时间的变化如图 3.4(c)所示,从图中可以看出,HMF 氧化为 FDCA 在 90 min 内基本完成,并且 FFCA 的浓度在整个测试过程中始终小于 1%。HMF 转化的法拉第效率为 98.6%(表 3.1)。值得注意的是,初始 HMF 溶液呈淡黄色,表明 HMF 在碱性条件下可以聚合或降解[10];但在氧化条件下,这些物质仍然可以被氧化成 FDCA，随着电化学氧化反应的进行，黄色逐渐褪色，如图 3.4(b)所示。然而，当阳极电势增加到了 1.43 V 时，HMF 转化的法拉第效率下降到了 77.2%（表 3.1 第 4 组），尽管反应可以更快地进行。法拉第效率的降低主要是因为水氧化反应的竞争。而当阳极电压降低到了在水氧化启动电势之下的 1.23 V，法拉第效率可以达到 99.4%，但是反应进行的比较慢，速率仅为 8.33×10^{-6} mmol·s^{-1}。这些结果表明，和目前已经报道的非贵金属催化剂相比，NiFe LDH 具有最高的反应活性和选择性。4 个连续周期的计时安培法测试，以评估该电极在 HMF 氧化期间的耐久性（图 3.4(d)）。4 个循环后，HMF 的转化率由 98% 下降到 93%，反应速率从 6.67×10^{-5} mmol-HMF·s^{-1} 下降到 6.20×10^{-5} mmol-HMF·s^{-1}。这种减少可能是由在相同的反应时间间隔内通过电极的电荷密度降低所致的，但每个循环的法拉第效率没有改变。

表 3.1 电化学氧化 HMF 的条件优化

序号	催化剂	C_{HMF} (mmol·L^{-1})	$E_{(onset)}$ V vs RHE	$E_{(j=20 mA·cm^{-2})}$ V vs RHE	电压 (V)	反应时间 (min)	HMF 转变	FDCA 产率	法拉第效率
1	NiFe LDH	0	1.37	1.53	—				
2	NiFe LDH	10	1.25	1.32	1.23	600	99%	98%	99.4%
3	NiFe LDH	10	1.25	1.32	1.33	90	98%	98%	98.6%
4	NiFe LDH	10	1.25	1.32	1.43	30	98%	97%	77.2%
5	NiFe LDH	50	1.18	1.30	1.33	350	92%	90%	98.7%
6	NiFe LDH	100	1.13	1.19	1.33	400	91%	90%	90.2%
7	NiAl LDH	0	1.55	1.96	—				
8	NiAl LDH	10	1.29	1.45					
10	NiGa LDH	0	1.55	1.70	—				
11	NiGa LDH	10	1.34	1.52					
12	$Ni(OH)_2$	10	1.28	1.41					
14	原始碳纤维纸	0	1.58	>2.0					
15	原始碳纤维纸	10	1.30	1.55					

将反应完成后的 NiFe LDH 催化剂进行 XPS 分析，其中的 Ni 2p 图谱显示出两个峰，分别位于 855.33 eV 和 856.50 eV（图 3.5(a)），可以归因于 $Ni(OH)_2$ 和 NiO。峰的形状和位置与反应之前的催化剂相比几乎没有变化，但是

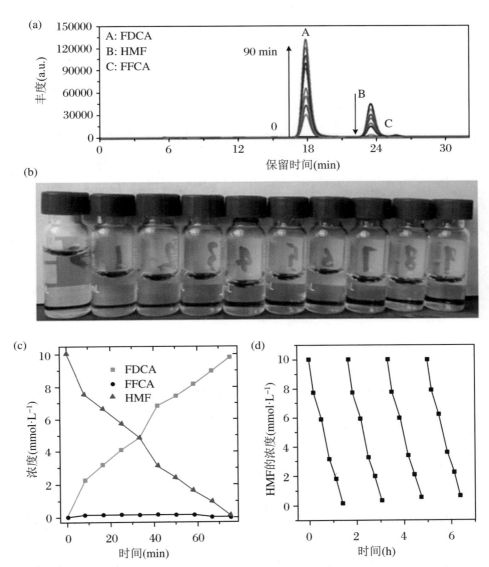

图 3.4 (a) 不同反应时间下不同产物的 HPLC 图谱（流动相溶液：1 mmol·L^{-1} H$_2$SO$_4$，流速：0.6 mL·min^{-1}）；(b) 电化学 HMF 氧化过程中阳极电解质的颜色变化；(c) 在 1.33 V·RHE^{-1} 时，HMF 及其氧化产物浓度随时间的变化；(d) HMF 浓度在 4 个连续周期内的变化情况

Ni(OH)$_2$ 的含量比反应前的有明显增加，表明在催化反应过程中有部分的 NiO 转化成了 Ni(OH)$_2$。Fe 2p 图谱（图 3.5(b)）表明在 HMF 氧化反应后有更多的 Fe^{3+} 生成。

NiFe LDH 纳米片电极在更高的 HMF 浓度下的电化学性能也得到了评价。从图 3.6(a) 的 LSV 曲线可以看出，随着 HMF 浓度的增加，E(onset) 和 E(j = 20 mA·cm^{-2}) 均减小（表 3.1 第 5～6 组）。如图 3.6(b) 所示，在所有 3

种 5-HMF 浓度下,5-HMF 浓度随时间的变化都可以拟合到一级动力学模型。在反应初始阶段,HMF 浓度为 50 mmol·L^{-1} 和 100 mmol·L^{-1} 时,对 FFCA 的选择性相对较高(图 3.6(c)和(d))。因此,相对于低浓度的 HMF 氧化,相应的法拉第效率略有下降。当 HMF 浓度超过 100 mmol·L^{-1} 时,HMF 从阳极到阴极的交叉变得更加显著。因此,需要改进膜组件来防止该现象。

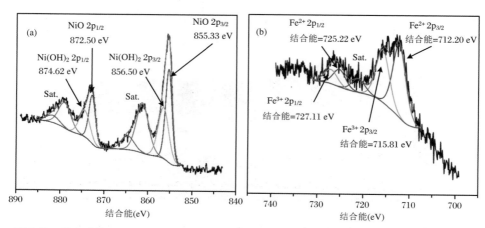

图 3.5　反应后的 NiFe LDH 电极 XPS Ni 2p(a)和 Fe 2p(b)图谱

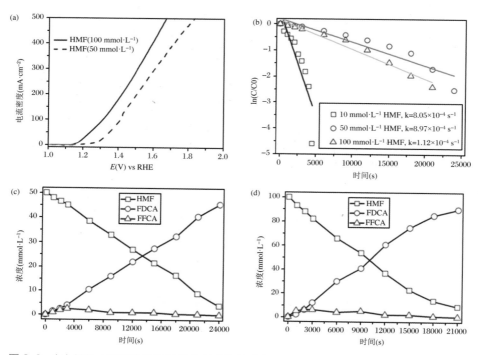

图 3.6　(a) NiFe-LDH/碳纤维纸催化剂在高浓度(50 mmol·L^{-1} 和 100 mmol·L^{-1}) KOH 中对 HMF 氧化的 LSV 曲线;(b) 不同浓度下 HMF 氧化的一级动力学模型;(c)和(d)分别为在计时安培试验期间 HMF 及其氧化产物的浓度变化(1 mmol·L^{-1} KOH 中电势 1.33 V vs RHE)

3.2 常温常压下零价金属对生物油的品质提升及其作用机制研究

3.2.1 概述

生物质热解过程中主要产物分成三相：固相的热解生物炭、液体产物生物油和少量的热解气体。其中的生物油是应用最为广泛的生物质热解产物，它既可以作为化石燃料的替代物，又可以作为化学原料提取高附加值的化学品[47-49]，故对它的研究也最为深入[50-58]。但是，由于生物油存在稳定性差、腐蚀性强、热值较低等缺点，其直接作为燃料或者化工原料的应用前景受到很大限制。因此，需要寻找合适的方法对生物油的品质进行提升，使其具有更加广泛的实用性。

生物油是一个多相混合物，包含大量的不同类型的化合物，例如水、烃类、含氧有机物、热解木质素等[59]。其中的一些化合物和生物油的缺点密切相关：生物油的强腐蚀性主要是由其大量存在的羧酸引起的[60-61]；而低热值主要是由大量的含氧化合物的存在导致的[48,62-63]；至于其化学不稳定性，则是由生物油中的醛酮类含有不饱和键的化合物容易发生聚合老化导致的[64-65]。降低这些化合物含量是提高生物油品质最主要的方法。目前常用的生物油品质提升方法主要有加氢[65-68]、催化裂解[69-70]和水蒸气重整等[71-72]。其中加氢是应用最为广泛的技术，其主要包括加氢脱氧和加氢酯化两种方法[67-68,73]。加氢脱氧主要是将不饱和醛酮类化合物转化为饱和的化合物（醇类或者是烃类等），可以大大降低生物油的含氧率，在提高生物油稳定性的同时还增加了其热值。而加氢酯化方法是将醛酮类先转化为醇类，再和生物油中的有机酸发生酯化反应，形成有机酸酯类，可以大大降低生物油中的有机酸含量，从而使其腐蚀性降低。可是这些过程，无论是加氢脱氧还是加氢酯化，都需要使用复杂的仪器设备和贵金属催化剂，并且在高温高压的条件下进行[73-75]。此外，高温高压的加氢过程通常会存在催化剂容易失活，反应器容易堵塞等各种问题[60,74]。为了避免这些问题的发生,寻找能在常温常压下实现生物油加氢是一个有意义的研究。

一些零价金属,包括铁、镁、铝、锌等具有很强的还原性,被广泛用于有机合成中的还原反应。例如,零价的铁和锌在硝基化合物的还原反应中是常用的还原剂,可以选择性地将硝基还原为氨基而不破坏其他基团[76-81]。而在Clemmensen还原反应中,零价锌是一个关键的还原剂用于将羰基转化为亚甲基[82-84]。所有这些反应都是在常温常压条件下、酸性体系中进行的,获得的产物的产率和选择性都很高。生物油体系也是一个酸性较强的体系,其中含有大量的醛酮类化合物,当零价金属加入到生物油体系中时,它可以和其中的有机酸首先发生反应,产生大量的活性氢自由基,这些活性氢自由基可以进一步和生物油中的不饱和醛酮类化合物反应,形成一些加氢饱和的产物(醇或者烃类)[85],可以显著提高生物油的稳定性和热值。此外,由于过程中的有机酸消耗,也可以大大降低生物油的腐蚀性。因此,使用活泼零价金属将可以有效地提升生物油的品质。更重要的是,反应是在常温常压的条件下进行的,并且没有额外的催化剂,保证了反应的易操作性和可持续性。

在本节研究中,我们首先采用不同的零价活泼金属对生物油的模型化合物进行加氢反应,通过活性测试选择出活性最好的金属用于实际的生物油体系,分析反应前后生物油的物理化学性质,以及所含化合物组成的变化,并以此阐述零价金属对生物油品质提升的可能作用机制。

3.2.2
零价金属催化生物油模型化合物的加氢反应及其机理

苯甲醛(分析纯)用作生物油中的醛酮类模型化合物,4种活泼金属(铁、铝、锌、镁)用作还原剂,都为粉末状,粒径小于 0.12 mm。乙酸作为生物油中的有机酸的模型化合物。生物油是通过流化床在 500 ℃ 热解稻壳而产生的,收集过来不经任何前处理,直接用于零价金属的加氢反应。首先将 5 mmol·L^{-1} 的零价金属粉末放置于 25 mL 的反应管中,随后加入 2.0 mmol·L^{-1}(0.212 g)的苯甲醛和 10 mmol·L^{-1}(0.58 g)的乙酸,以及 3.0 g 有机溶剂和 1.0 g 水。然后在常温常压条件下,将混合物置于磁力搅拌器中搅拌一段时间使其反应。反应完成之后获得的混合物使用针孔滤膜进行过滤。反应物的转化率和产物的选择性通过气相色谱进行测定。色谱条件如下所示:色谱柱为毛细管柱(规格:1.0 μm,50 m 和 0.25 mm);检测器:氢火焰离子化检测器,使用萘作为内标物;

进样器和检测器的温度均为 250 ℃,柱温从 60 ℃以 8 ℃·min^{-1}的速率升高至 230 ℃,保持 2 min;反应物(苯甲醛)和产物(苯甲醇、甲苯)的检测是通过和相应标准品进行对比得到的,其含量通过相应色谱峰的面积计算而得,苯甲醛的转化率(Conversion(%))和苯甲醇的选择性(Selectivity(%))分别通过以下方程计算而得

$$转化率(\%) = \frac{C_{r0} - C_{rt}}{C_{r0}} \times 100\% \tag{3.1}$$

$$选择性(\%) = \frac{C_{pt}}{C_{r0} - C_{rt}} \times 100\% \tag{3.2}$$

其中,C_{r0}是初始苯甲醛的浓度(mmol),C_{rt}是在 t 时刻苯甲醛的浓度(mmol),C_{pt}是在 t 时刻苯甲醇的浓度。

苯甲醛是一个典型的具有高反应活性和化学不稳定性的羰基化合物[86-87],被选作生物油的模型化合物用于测试不同零价金属的加氢活性。乙酸则作为生物油中存在最多的有机酸,用来模拟生物油中的强酸性环境。4 种不同的活泼零价金属用作加氢还原剂测试其对苯甲醛的加氢活性。表 3.2 所示的是不同零价活泼金属在常温常压条件下对苯甲醛的加氢反应性能。由表可知,零价金属锌的反应活性最高,在其作用下,苯甲醛的转化率大于 95%,几乎可以定量地转化为苯甲醇(选择性大于 99%)。这表明无论是苯甲醛的转化率和苯甲醇的选择性都可以和高温高压条件下氢气作为氢源的加氢反应结果相比[88-90]。

表 3.2 不同零价活泼金属对生物油模型化合物苯甲醛的加氢反应

序号	溶剂①	金属	有机酸	反应时间(min)	苯甲醛转化率	苯甲醇选择性	甲苯选择性
1	THF	Al	CH_3COOH	180②	痕量	痕量	痕量
2	THF	Fe	CH_3COOH	180②	痕量	痕量	痕量
3	THF	Mg	CH_3COOH	30③	11%	87%	13%
4	THF	Zn	CH_3COOH	60	95%	>99%	<1%
5	CH_3OH	Zn	CH_3COOH	60	96%	>99%	<1%
6	CH_2Cl_2	Zn	CH_3COOH	60	95%	>99%	<1%

注:① 所用溶剂是表中所列有机溶剂和水以 3∶1(质量比)混合而成的。
② 由于铁和铝的反应比较慢,因此延长其反应时间至 180 min。
③ 零价金属镁和乙酸的反应非常快,在 30 min 之内金属镁基本完全反应。

根据以上结果，我们推测零价金属锌参与的对生物油模型化合物的加氢过程机理如图 3.7 所示。首先金属锌吸附有机酸电离产生的水合质子，然后将其电子转移至质子上，使得质子转化为原子状态的氢，这是一个具有高度还原活性的自由基，其半衰期只有 0.3 s。活泼原子氢自由基迅速将其电子转移给其他的电子受体（苯甲醛的羰基），其本身也结合在羰基之上，形成了加氢的产物[91]。在铝和铁的作用下，经过 180 min 的反应，苯甲醛都基本不能转化为苯甲醇和甲苯等氢化产物，这主要是因为这两个金属和有机酸的反应比较慢，产生不了足够的活泼原子氢进行氢化反应。而对于金属镁，由于它和有机酸的反应十分剧烈，产生的活泼氢原子过于迅速，大部分都直接相互结合形成氢气了，氢气在常温常压条件下十分稳定，不能有效参与对羰基化合物的氢化作用[86-87]。

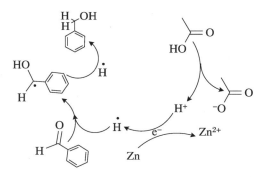

图 3.7　零价金属锌参与的苯甲醛的氢化作用机理示意图

图 3.8 所示的是不同反应时间下苯甲醛的转化率以及苯甲醇和甲苯的产率，从图中可知，苯甲醛的氢化反应是一个十分迅速的过程，高于 85% 的苯甲醛可以在反应开始的 20 min 内发生转化，并且几乎定量地生成苯甲醇；对于甲苯，作为苯甲醛加氢脱氧的产物，其产率在 10 h 的反应时间内始终低于 1%，表明在零价金属参与的苯甲醛的加氢过程中，脱氧的反应是很难进行的。根据以上分析可知，零价金属锌在对生物油模型化合物的氢化反应中表现出最高的反应活性，并且反应过程十分迅速，产物转化彻底，选择性较高。因此，在接下来对生物油实际体系的氢化反应中，我们选择零价金属锌作为还原剂。

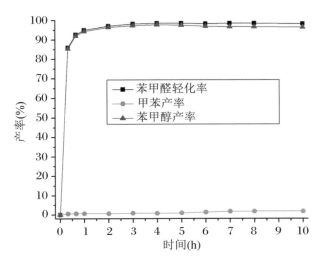

图 3.8　不同反应时间下苯甲醛的转化率以及苯甲醇和甲苯的产率

3.2.3
生物油品质提升前后的性质分析

将 0.50 g 的零价金属粉末、10 g 的生物油和 2.0 g 的四氢呋喃(THF)加入 100 mL 圆底烧瓶中,置于磁力搅拌器中,在常温常压的条件下搅拌 300 min。反应结束之后的混合物通过针孔滤膜进行过滤,滤液和固体残渣储存于冰箱中备用。加氢反应前后生物油的元素组成(C,H,N,O)通过元素分析仪进行测定。锌的含量是通过化学消解之后再采用原子发射光谱(ICP-AES,Optima 7300 DV,Perkin Elmer Co.,USA)进行测定的。反应前后生物油的 pH 通过 pH 计来测定。反应前后生物油中的热解木质素的含量通过文献报道的方法进行分离和测定[92]:将 100 g 的生物油缓慢滴加至 1000 mL 的冰氯化钠溶液(0.1 mol·L^{-1})中使得其中的热解木质素析出,然后将析出的热解木质素过滤分离,重新分散悬浮于冰水中搅拌 240 min 洗去其中的可溶性物质,最后再抽滤使其分离,然后将所得的固体残渣在 30 ℃下真空干燥,将获得干燥过的热解木质素称重,计算其产率。

反应前后生物油的核磁共振碳谱通过一个 400 MHz 的核磁共振波谱仪(AV 400,Bruker Inc.,Switzerland)在常温下表征。对于 GC-MS 分析,测定之前,生物油先通过二氯甲烷进行萃取,获得的有机相通过 GC-MS 分析其主要成分的结构。GC-MS 的仪器是由气相色谱（Agilent-7890A）和质谱（Agilent-975C）组成。使用 HP-5MS 非极性毛细管柱（30 m-0.25 mm-0.25 μm）。使用 1.0 mL·min^{-1} 的高纯氦气作为载气，分流比为 1∶20。测试过程采取程序升温的模式，毛细管柱温首先以 4 ℃·min^{-1} 从 40 ℃ 升高至 180 ℃，然后以 10 ℃·min^{-1} 升高至 250 ℃，在这个温度下保持 5 min，然后开始降温。检测器的温度保持 250 ℃ 不变。

图 3.9 所示的是生物油品质提升前后的光学照片，从中可知，初始生物油是一种黑色不透明的液体，而经过和零价金属的反应之后，变成了深棕色的透明液体。表 3.3 所示的是反应前后生物油的元素组成和主要物理化学性质的比较，从中可知，和零价金属锌反应后，生物油中的氧含量显著降低，碳和氢的含量明显的上升，相应地，其热值也有一定程度的提高（从反应前的 12.5 MJ·kg^{-1} 提高到反应后的 13.4 MJ·kg^{-1}）。而热解木质素，作为一个影响生物油的稳定性的主要成分[93]，其含量从反应前的 5.83% 下降到反应后的 4.62%，证明生物油的稳定性有了一定的提高。生物油的 pH 从反应前的 3.53 提高到了 4.85，反映了生物油中的有机酸的含量有了很大的降低，这也就使得生物油的腐蚀性有了较大的减弱。以上这些结果表明，和零价金属锌反应后，生物油的品质，包括热值、稳定性、腐蚀性等都有了较大的提升。

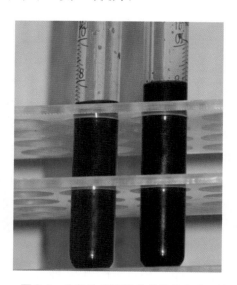

图 3.9 生物油品质提升前后的光学照片

根据在整个反应过程中的锌元素的质量平衡方程，我们计算出锌元素在反应之后在不同相的分布情况：

$$Zn_{add} = Zn_{生物油} + Zn_{固体残渣} \quad (3.3)$$

其中，$Zn_{add} = 0.5$ g 是反应中加入的金属锌的质量；$Zn_{生物油} = 69.0$ mg·kg^{-1} 是

锌元素在反应后生物油中的含量；$Zn_{固体残渣}$是锌元素在反应后的固体残渣中的含量。

结果表明，反应过后，多达99.8%的锌元素仍然集中在固体残渣之中，这部分的锌元素很容易回收再利用。但是和初始生物油相比，反应后生物油中的锌元素含量达到了69.0 mg·kg^{-1}，仍然在一个较高的水平。考虑到锌离子的亲水性，它应该主要存在于生物油的水相之中。我们对反应过后的生物油进行了一个相分离过程，并且测定了其中的有机相中的锌元素含量，发现其含量迅速降低到了2.6 mg·kg^{-1}。这个结果表明，提质后的生物油并没有被锌污染，可以作为燃料或者化工原料安全使用。

表3.3 生物油和零价金属锌反应前后的元素组成和物理化学性质的比较

	反应前	反应后
C(wt%)	34.1	36.4
H(wt%)	9.0	9.5
N(wt%)	0.2	0.2
O(wt%)	50.6	48.4
Zn(mg·kg^{-1})	n.d.[①]	69.0(2.6)[②]
热值(MJ·kg^{-1})[③]	12.5	13.4
热解木质素(wt%)	5.83	4.62
含水率	46.3	42.5
pH	3.53	4.85

注：① 未检测到。

② 69.0 mg·kg^{-1}是全体系生物油中的锌含量，而2.6 mg·kg^{-1}是锌在生物油中经过有机物萃取得到的有机相中的含量。

③ 热值(MJ·kg^{-1}) = (3.55 C^2 − 232 C − 2230 H + 51.2 $C×H$ + 131 N + 20600)/1000[62]，见公式(2.17)。

生物油提质前后的化学特征变化通过^{13}C核磁共振波谱进行进一步的表征。表3.4所示的是根据核磁共振波谱图解析出来的不同化学特征的碳原子的含量[94-96]。反应前后核磁结果最明显的变化就是羰基(C=O，化学位移为170~215 mg·kg^{-1})的含量显著的降低(从9.8 mol%降低至3.1 mol%)，同时伴随着脂肪族碳氧单键(C—O，化学位移为50~90 mg·kg^{-1})的含量明显的增加(从36.8 mol%升高至44.3 mol%)，证明了在反应过程中C=O被金属锌转化为C—O的过程。

表 3.4　生物油和金属锌反应前后核磁共振碳谱分析结果比较

化学位移 (mg·kg^{-1})	碳原子的化学形态	含量*	
		反应前	反应后
0～50	C—C(脂肪类)	13.9	16.9
50～90	C—O(醇和醚类)	36.8	44.3
90～120	C—C(芳香类)和C═C(烯烃)	26.8	23.5
120～150	C—O(酚类和芳香醚类)	12.7	12.2
170～215	C═O(醛,酮和羧酸类)	9.8	3.1

注：*总的碳原子的摩尔比。

3.2.4
零价金属锌对生物油品质提升过程机制探讨

首先我们对生物油品质提升前后所含化合物的分子结构和相对含量的变化进行了 GC-MS 表征。图 4.10 所示它们的 GC-MS 图谱：从中可以看出最明显的变化是反应后的生物油 GC-MS 图谱中发现了 13 个新的峰(编号为 S1～S13)。我们对 GC-MS 中的主要的峰进行了分子结构的确认，结果如表 3.5 所示：乙酸是在初始生物油中含量最高的组分，但是经过和零价金属锌的反应之后，其含量大大降低了，其他的有机酸类，包括 2-羟基丙酸(2-hydroxypropanoic acid)和 2,3-二羟基丙酸(2,3-dihydroxypropanoic acid)的含量也显著地降低了。有机酸的含量的降低是导致生物油的 pH 升高的主要原因，从而可以显著降低生物油的腐蚀性。另一个值得注意的变化就是醛类化合物的含量在反应后的生物油中显著降低：糠醛在初始生物油中的相对含量居第三，达到 11.3%，经过和金属锌的反应，其相对含量降低到了 2.8%，其他的醛类，例如丁二醛(succinaldehyde)、4-羟基-3-甲氧基-苯甲醛(4-hydroxy-3-methoxybenzaldehyde)和 4-羟基-3,5-二甲氧基-苯甲醛(4-hydroxy-3,5-dimethoxybenzaldehyde)等的含量都有一定程度的降低。

表 3.5 反应前后生物油中主要化合物的分子结构和相对含量比较

序号	保留时间（min）	分子式	分子结构[①]	相对含量(%) 反应前	相对含量(%) 反应后
1	3.989	$C_2H_4O_2$	acetic acid	12.2	3.7
2	4.852	$C_3H_6O_2$	4-hydroxybutan-2-one	13.7	10.2
3	5.681	$C_4H_8O_3$	ethyl-2-hydroxyacetate	3.4	2.6
4	6.025	$C_3H_6O_3$	2-hydroxypropanoic acid	1.7	0.5
5	8.082	$C_4H_8O_2$	1-hydroxybutan-2-one	3.1	2.9
6	8.674	$C_4H_6O_2$	succinaldehyde	2.2	0.3
7	9.102	$C_3H_6O_4$	2,3-dihydroxypropanoic acid	1.7	0.7
8	10.763	C_5H_4O	furfural	11.3	2.8
9	12.322	$C_6H_{10}O_3$	3-oxobutyl acetate	0.6	1.0
10	13.510	$C_6H_8O_3$	2,5-dimethoxyfuran	1.1	0.3
11	13.952	$C_5H_6O_2$	3-methylfuran-2(5H)-one	0.3	0.3
12	14.140	$C_4H_4O_2$	furan-2(3H)-one	3.9	4.1
13	14.935	$C_6H_{10}O_2$	3-methylpentane-2,4-dione	0.3	n.d.[②]
14	15.288	$C_5H_6O_2$	2-methoxyfuran	0.6	0.7
15	16.515	C_6H_8O	2,5-dimethylfuran	1.1	1.4
16	16.914	$C_5H_6O_2$	4-methylfuran-2(5H)-one	1.0	1.1
17	17.169	C_6H_6O	phenol	2.1	2.3
18	17.532	—	cannot beidentified	1.4	1.3
19	17.993	—	cannot beidentified	2.7	2.6
20	18.302	$C_6H_8O_2$	1,3-dimethoxycyclopentane	0.4	0.5
21	19.093	$C_7H_{14}O_2$	3-methylcyclopentane-1,2-dione	3.9	5.1
22	19.947	$C_7H_{12}O_2$	5-isopropyldihydrofuran-3(2H)-one	0.6	0.5
23	20.225	C_7H_8O	o-cresol	1.4	1.6
24	21.064	C_7H_8O	p-cresol	1.6	1.6
25	21.595	$C_7H_8O_2$	2-methoxyphenol	4.7	4.9
26	22.356	$C_6H_{10}O_3$	4-hydroxy-4-methyltetrahydro-2H-pyran-2-one	0.6	0.4
27	22.853	$C_8H_{10}O_2$	2-methylbenzene-1,3-diol	0.3	0.3
28	24.560	$C_8H_{10}O$	4-ethylphenol	3.1	3.3
29	25.502	$C_8H_{10}O_2$	2-methoxy-4-methylphenol	4.7	4.5

续表

序号	保留时间 (min)	分子式	分子结构[①]	相对含量(%) 反应前	相对含量(%) 反应后
30	25.742	$C_4H_6O_3$	4-methyl-1,3-dioxolan-2-one	0.6	0.4
31	26.987	—	cannot beidentified	0.3	n.d.
32	27.873	—	cannot beidentified	0.3	n.d
33	28.499	$C_9H_{12}O_2$	4-ethyl-2-methoxyphenol	0.8	0.7
34	30.192	$C_8H_{10}O_2$	2,6-dimethoxyphenol	0.8	0.6
35	32.534	$C_8H_8O_3$	4-hydroxy-3-methoxybenzaldehyde	1.7	n.d
36	32.694	$C_8H_{12}O_3$	2-hydroxy-6-methylcyclohex-3-enecarboxylic acid	0.6	n.d
37	34.313	—	cannot be identified	0.5	0.7
38	35.209	$C_{10}H_{12}O_3$	1-(4-hydroxy-3-methoxyphenyl)propan-2-one	1.0	0.8
39	35.991	—	cannot be identified	0.5	0.8
40	36.434	$C_8H_8O_4$	1-(2,3,4-trihydroxyphenyl)ethanone	1.2	1.0
41	39.851	$C_{10}H_{14}O_3$	2-ethoxy-6-(methoxymethyl)phenol	0.4	1.6
42	40.116	$C_9H_{10}O_3$	4-hydroxy-3,5-dimethoxybenzaldehyde	0.5	n.d.
43	41.923	$C_{10}H_{10}O_3$	3-(4-hydroxy-2-methoxyphenyl)acrylaldehyde	0.5	0.3

注：① 新生成的化合物的结构和相对含量其他表列出。

② n.d. 代表相对含量小于0.3%。

新生成的13种化合物的结构确认和相对含量变化，以及可能的形成途径如表3.6所示。化合物S1~S4以及S12主要是相应的醛酮类化合物氢化生成的醇类物质。化合物S5~S7是酯类化合物，主要来源于有机酸和醇类的酯化反应。其他新生成化合物(S8~S11,S13)都是酚类化合物，它们被认为是来自热解木质素的还原碎片[97-98]，热解木质素是植物的木质素化合物在热解过程中形成的寡聚体[99-100]，在活泼氢的存在下，热解木质素中的甲氧基苯丙烷结构会发生破坏，一些芳香醚键将被氢化，形成芳香类的羟基化合物。此外，反应过程形成的Zn(Ⅱ)-羧酸配位化合物也可以作为原位催化剂催化芳香醚键的氢化作用[101]。化合物S8~S13的生成也可以解释为什么反应过后生物油中的热解木质素含量会显著地下降。

表 3.6　在生物油和零价金属锌反应过程中新生成的 13 种化合物的结构，相对含量及可能的形成途径

序号	化合物结构式	相对含量	可能的形成途径
S1	pentan-1-ol	0.5%	peak 2 → pentan-1-ol
S2	butane-1,3-diol	0.7%	peak 2 → butane-1,3-diol
S3	butane-1,4-diol	0.4%	peak 6 → butane-1,4-diol
S4	furan-2-ylmethanol	3.1%	peak 8 → furan-2-ylmethanol
S5	ethyl-2-acetoxyacetate	1.0%	peak 1 + peak 3 → ethyl-2-acetoxyacetate
S6	furan-2-ylmethyl acetate	1.4%	乙酸 + furfuryl alcohol → furan-2-ylmethyl acetate
S7	2-hydroxybutyl-2-hydroxypropanoate	0.8%	peak 4 + 二醇 → 2-hydroxybutyl-2-hydroxypropanoate
S8	2-methoxyphenol	0.4%	热解木质素的碎片还原
S9	2-methoxy-3-(2-propenyl)phenol	0.8%	热解木质素的碎片还原

续表

序号	化合物结构式	相对含量	可能的形成途径
S10	3,4-dihydroxybenzoic acid	0.3%	热解木质素的碎片还原
S11	2,6-dimethoxy-4-vinylphenol	0.6%	热解木质素的碎片还原
S12	4-(hydroxymethyl)-2,6-dimethoxyphenol	0.3%	peak 42 → (结构式)
S13	2,4,6-trimethoxyphenol	0.9%	热解木质素的碎片还原

在本节研究中,我们采用零价活泼金属在常温常压下对生物油进行品质提升,使其更好地应用于其他领域。根据相关的实验结果,可以得出以下结论:

(1) 对于生物油的模型化合物的加氢过程,金属锌表现出最好的性能,可以几乎定量地将苯甲醛转化为苯甲醇,选择性大于 99%。锌对苯甲醛的氢化反应主要机理是锌先和有机酸反应生成活泼原子氢,再由活泼原子氢对苯甲醛的羰基进行氢化,生成相应的氢化产物苯甲醇。

(2) 在实际的生物油体系中,金属锌也表现出很好的反应性能,它可以显著地提高反应后生物油的性质,包括降低腐蚀性、提高稳定性和热值等。

(3) 反应后生物油的物化性质以及主要化合物的结构和相对含量也发生了明显改变。新生成了 13 种可检测化合物,其主要生成路径包括醛酮类的直接氢化,生成的醇类和生物油中有机酸的酯化,以及热解木质素碎片的氢化反应等。

3.3 快速热解铜负载生物质选择性提高生物油的品质

3.3.1 概述

很多重金属元素(铜、锌、镍等)通常具有较强的催化活性,并且广泛应用于催化各类化学反应,例如有机合成、污染物降解以及催化重整合成等[102-107]。含有这些重金属元素的催化也同时在生物质的热化学转化生成燃料或者化学品的过程中有大量的应用[108-110]。然而,制备这些催化剂通常涉及较为复杂的物理化学过程,或者需要用到大量的有毒有害试剂[111-113]。考虑到生物质材料本身具有丰富的功能基团,可以直接从废水中吸附这些重金属[114-118],达到去除水体中重金属污染的目的,同时吸附后的生物质通常被这些重金属污染而难以直接处置。因此,如果能利用重金属的催化作用,使其在生物质热解过程中起到原位催化作用,从而选择性地提高液体产物生物油的产率和品质是一个很有前景的研究。然而,相关的研究尚未见文献报道。

因此,在本研究中,我们选择了金属铜作为模型重金属,将其通过生物吸附过程负载于木屑生物质上,然后将所得的铜负载的木屑生物质在不同温度条件下热解,分析产生的生物油的物理化学性质以及所含化合物的组成,并且和未经铜负载生物质热解产生的生物油进行对比,研究铜在热解过程中对所产生生物油品质提升的作用机制。同时,金属铜在热解过程中的迁移转化过程也通过XRD、XPS等分析表征手段进行了深入的研究。

3.3.2
铜催化生物质热解过程及影响因素分析

木屑生物质(Fir Sawdust,FSD)收集自当地的一个木材处理厂,使用之前,木屑先用水清洗几次以去除其中的灰尘颗粒或者其他杂质,然后烘干并用高速粉碎机粉碎、过筛,收集粒径小于 0.12 mm 的颗粒。木屑生物质的近似分析以及元素组成分别通过热重分析仪和元素分析仪进行测定,结果如表 3.7 所示。分析纯的硫酸铜($CuSO_4$)用作铜源配制含铜的废水。

表 3.7 木屑生物质的近似分析和主要元素组成

检测项	近似分析(wt%)	元素	组成(wt%)
含水率	3.32	C	47.30
挥发性物质	79.40	H	5.69
固定碳	15.65	N	0.23
灰分	1.63	O[①]	46.78
		S[②]	1.87×10^{-3}

注:① 通过差值法计算($O\% = 100\% - C\% - H\% - N\%$)。
② 通过 ICP-AES 测定。

铜负载的生物质(Cu-FSD)是使用木屑生物质作为吸附剂,从废水中吸附重金属铜得到的。吸附在常温常压的条件下进行,一定量的木屑生物质和 100 mL 的含铜废水,在 200 r·min^{-1} 搅拌情况下吸附 300 min 之后,将所得混合物中的水分挥发完全,即获得 Cu-FSD。

生物油和热解生物炭的主要元素组成通过元素分析仪进行测定,木屑生物质的热化学特性通过热重分析仪进行表征。Cu-FSD、生物油和热解生物炭中的铜含量先通过化学消解,再用原子发射光谱进行测定。XPS 图谱用于分析生物质热解前后铜的化学状态的变化情况,通过一个 X 射线光电子能谱仪来测定。XRD 图谱用于分析热解生物炭中铜的晶型变化情况,测定仪器(ESCAL-AB250,Thermo-VG,Scientific Inc.,UK)为 18 kW 旋转阴极 X 射线衍射仪(MXPAHF,Japanese Make Co.,Japan),测定角度范围为 20°~80°,扫描速率为 0.02°·s^{-1}。生物油的紫外光谱通过一个紫外可见分光光度计(UV-1700,Phenix Co.,Ltd.,China)进行测定。测定之前,准确称量 50 mg 的生物油溶解于无水乙醇中,并定量至 50 mL,然后置于紫外可见分光光度计中,在波长范围为 200～600 nm 区间进行扫描得到生物油的紫外可见吸收光谱。生物油的核磁共振碳谱通过 400 MHz 的核磁共振波谱仪（AV 400，Bruker Inc.，Switzerland）在常温下表征。测试之前，准确称量 50 mg 的生物油样品，溶解于 1 mL 的氘代丙酮中，然后装入专用核磁管中进行核磁测定。生物油中组分的化学结构通过气相色谱-质谱联用仪（GC-MS）测定。测定之前，生物油先通过二氯甲烷进行萃取，获得的有机相通过 GC-MS 分析其主要成分的结构。GC-MS 的仪器是由气相色谱（Agilent-7890A）和质谱（Agilent-975C）组成，使用 HP-5MS 非极性毛细管柱（30 m-0.25 mm-0.25 μm）。使用 1.0 mL·min^{-1} 的高纯氦气作为载气，分流比为 1∶20。测试过程采取程序升温的模式，毛细管柱温首先以 4 ℃·min^{-1} 从 40 ℃升高至 180 ℃，然后以 10 ℃·min^{-1}升高至 250 ℃，在这个温度下保持 5 min，然后开始降温。检测器的温度保持 250 ℃不变。

金属铜的负载量和热解温度是影响 Cu-FSD 的热解行为的主要因素。首先，我们研究了不同铜的负载量对 Cu-FSD 热解过程中生物油的产率和性质的影响。图 3.10(a)所示的是不同负载量条件下 Cu-FSD 热解产物（生物油、热解生物炭和热解气）的产率分布，从中可知，Cu-FSD 热解生物炭的产率明显高于 FSD，并且当铜含量达到 0.5%时，其产率最高。而 Cu-FSD 的热解气体产率则大大低于 FSD，并且其产率受铜的负载量的影响比较小。Cu-FSD 的低热解气体产率主要原因可能是由于在铜的催化作用下，热解气体（氢气和小分子烃类）组分会进一步和生物油和热解生物炭反应，生成可冷凝的生物油组分，例如，铜可以催化氢气和热解木质素之间的反应生成一些小分子的酚类化合物[97]，铜的存在也可能催化小分子烃类的气相芳香化反应，形成可冷凝的芳香烃类存在于生物油相中[119]，从而导致了热解气体产率的降低。生物油的产率从铜的负载量为 0 时的 47.3%上升到铜负载量为 1.0%时的 54.3%，随着铜的负载量进一步上升，生物油的产率不断下降，当铜的负载量达到 2.5%以上时，生物油的产率甚至低于未经铜负载生物质热解的生物油产率。这个趋势和文献报道的玉米芯的催化热解过程中生物油产率的变化是相似的[120]。生物油的热值和其产率的变化趋势基本一致，生物油热值的最大值为 14.79 MJ·kg^{-1}，是在铜负载量为 1.0%时所产生物油中测得的。而当铜的负载量达到 2.5%和 3.0%时，生物油的热值分别降低到了 11.02 MJ·kg^{-1}和 11.21 MJ·kg^{-1}，甚至低于初始生物质所产生物油的热值。因此，理想的铜负载量为 1.0%，在这个铜含量下，热解产生的生物油的热值和产率都是最佳的。这个现象可以解释为：在低的铜含量情况下（<1%），铜的存在可以促进热解过程中的脱氧反应[121]，从而使得产生的生物油氧含量降低，提高其热值。而随着铜含量的上升，铜的催化脱氧能力受到其

氧化作用的影响,而作为一个较强的氧化剂(Cu(Ⅱ) + 2e$^-$ = Cu(0),E_0 = 0.34 V),大量铜的存在可以导致生物质在热解过程中产生的挥发组分发生一定的氧化,从而使其热值降低。

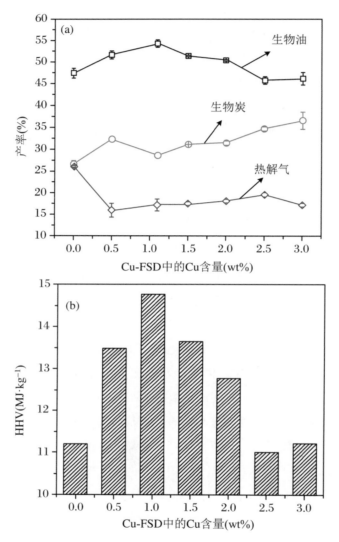

图 3.10　不同铜含量的 Cu-FSD 热解产物的产率分布(a)
以及所得生物油的热值变化(b)

热解温度为 500 ℃

温度是另一个影响 Cu-FSD 热解过程的主要因素,它不仅可以影响热解产物的产率分布,还对产物的化学性质和组成有明显的影响[95,122]。图 3.11(a)所示的是不同温度下 Cu-FSD 和 FSD 热解产物的产率分布,从中可知,热解生物炭和热解气体的产率随着温度的升高分别呈现出单调下降和上升的趋势,而生物油的产率则是先上升后下降,在温度为 500 ℃ 时达到最大值。这主要是因为在低的热解温度下,生物质中的挥发组分不能完全分解挥发出来,温度的升高有利于生物质的分解产生挥发组分,冷凝而成生物油;但是当温度达到一定值时,再升高温度尽管可以增加生物质的分解产生挥发组分,同时已经产生的挥发组分也将继续裂解,产生小分子不可冷凝气体,导致生物油产率的降低。因此,存在一个最佳温度值,既可以保证生物质的充分分解产生挥发性物质,又能尽可能地抑制挥发性物质的再次裂解,从而使生物油的产率达到最大值[120,123]。此外,值得注意的是,Cu-FSD 热解生物油产率在所有的热解温度下都显著高于 FSD,表明铜的存在确实可以提高生物质热解过程中生物油的产率。图 3.11(b)所示的不同温度下所得生物油的热值,从中可知,FSD 热解所产生物油的热值在 $10.77\sim12.32$ MJ·kg^{-1} 之间,明显低于 Cu-FSD 热解所产生物油。这主要是因为一定含量的铜的存在可以降低生物油中的氧元素含量(表 3.8),从而提高生物油的热值。以上结果表明,一定含量的铜的存在可以显著改善生物质热解过程中的生物油的热值和产率。

表 3.8　不同铜含量条件下 Cu-FSD 和 FSD 热解所产的生物油的元素组成和含水率的比较

铜含量 (wt%)	元素组成(wt%)				含水率
	C	H	N	O	
0	30.25	8.32	0.24	61.19	37.12
0.50	34.64	7.44	0.36	57.56	28.38
1.09	37.75	7.20	0.21	54.88	31.63
1.50	35.01	7.37	0.37	57.25	33.21
2.00	32.41	7.20	0.54	59.85	39.09
2.50	29.91	8.38	0.31	61.40	52.18
3.00	29.58	7.95	0.38	62.09	52.47

3.3.3

金属铜对生物油的化学组成和性质的影响及其作用机制

图 3.12 所示的是 Cu-FSD 和 FSD 热解所产生物油的紫外-可见吸收光谱的比较。由图可见,每条谱线都有两个吸收峰分别位于波长为 206 nm 和 270 nm

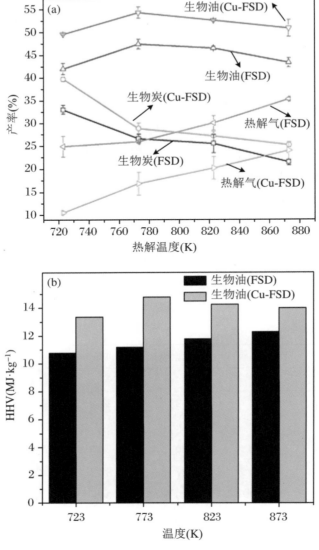

图 3.11 （a）不同温度 Cu-FSD 热解产物的产率分布；
（b）不同温度 Cu-FSD 热解所得生物油的热
值变化（铜负载含量 1%）

处。这两个吸收峰是苯环的 E2 和 B 带的紫外吸收，是表征芳香化合物的特征紫外吸收。比较 Cu-FSD 和 FSD 生物油的紫外吸收谱线发现，这两个吸收峰的位置基本相同，不同的只是紫外吸收强度，Cu-FSD 生物油的 E2 谱带吸收强度为 1.65，大于 FSD 生物油的 1.49，B 谱带的紫外吸收也有相同的规律。以上结果表明，Cu-FSD 生物油含有更多的芳香化合物，这与之前推测的铜的存在能促进生物质热解过程芳香化合物的形成是一致的。

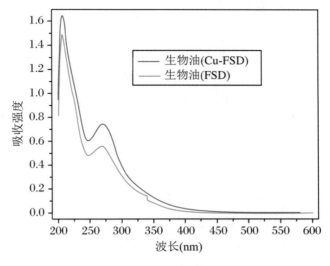

图 3.12 Cu-FSD 和 FSD 热解生物油的紫外光谱分析

^{1}H-核磁共振(^{1}H-NMR)用于分析生物油中氢原子的不同化学状态,从而推测出生物油中所含主要化合物的基本性质,结果如表 3.9 所示。Cu-FSD 所产生物油中除了水合羟基中的氢原子小于 FSD 所产生物油,其他化学环境下的氢原子含量均比 FSD 多。最主要的是,Cu-FSD 生物油中的芳香环上的氢原子含量达到 6.84 mol%,几乎是 FSD 生物油的两倍,而芳香化合物的主要来源包括两个方面,其一是木质素的分解生成[97,124],其二是热解所产小分子气体烃类的进一步反应生成。Cu-FSD 中芳香化合物的增加和其紫外可见吸收光谱的结果是一致的,进一步证实铜的存在可以促进热解过程中芳香化合物的生成。

表 3.9 Cu-FSD 和 FSD 热解所产生物油的 ^{1}H 核磁共振结果分析

化学位移 (mg·kg^{-1})	功能基团	含量(mol%)*	
		FSD	Cu-FSD
0.5~3.0	脂肪基(H—C—R)	37.62	41.46
3.0~3.4	甲氧基(H$_3$C—O—R)	3.70	3.91
3.4~3.7	水(H$_2$O)	48.01	37.34
3.7~5.5	羟基(H—O—R)	6.07	8.51
6.0~8.2	芳香基(H—⬡—R)	3.47	6.84
9.2~10.0	醛基(HC(=O)—R)	0.10	1.19

注:* 相对 H 原子的含量。

GC-MS 用于分析生物油中主要化合物的分子结构及其相对含量,结果列于表 3.10。从中可知,无论是 Cu-FSD 还是 FSD,其热解所产生物油都含有一系列复杂的含氧化合物,包括羧酸、醛、酮、醚和酚类,这与文献报道的其他来源的生物油的组成是一致的[125-126]。和 FSD 热解生物油相比,Cu-FSD 热解生物油的化合物组成是相似的。但是,所有化合物的相对含量变化是比较明显的。由于生物油中的化合物基本上都有两种及以上的功能基团,因此难以通过功能基团将其很好地分类。我们将生物油中的化合物通过其所含碳原子数进行分类,例如 C3 代表分子中含有 3 个碳原子的化合物。图 3.13 所示的是不同碳原子数的化合物在 Cu-FSD 和 FSD 热解生物油中的分布情况。其中 C7~C10 化合物被认为主要是生物质中的木质素热解产生的,是生物油中的主要成分,分别占 FSD 生物油的 66.5%和 Cu-FSD 生物油的 80.1%。含碳原子个数大于 11 的化合物(C11+)被认为是木质素不完全热解产生的碎片,在 Cu-FSD 生物油中的含量远远低于其在 FSD 生物油中的含量。以上结果表明,铜的存在可以有效地促进生物质中木质素的热解产生小分子芳香化合物,这和之前的核磁共振波谱和紫外可见光谱分析结果是一致的。

表 3.10 Cu-FSD 和 FSD 热解生物油的主要成分的分子结构和相对含量比较

时间(min)	分子式	分子结构	相对含量	
			FSD 生物油	Cu-FSD 生物油
2.27	$C_2H_4O_2$	acetic acid	5.31	4.39
2.41	$C_4H_8O_2$	3-hydroxylbutanal	3.30	1.29
2.54	$C_3H_6O_2$	1-hydroxypropan-2-one	2.79	0.72
3.98	$C_5H_{10}O_2$	1-hydroxyl-3-methylbutan-2-one	0.16	0.23
4.59	$C_4H_6O_3$	4-methyl-1,3-dioxolan-2-one	0.29	0.37
4.93	$C_3H_4O_3$	1,3-dioxolan-2-one	n. d.*	0.18
5.04	$C_5H_{10}O_2$	Cyclopentane-1,3-diol	1.14	1.02
6.58	$C_7H_{12}O_2$	3,4,4-trimethyldihydrofuran-2(3H)-one	0.80	n. d.
6.78	C_5H_6O	cyclopent-2-enone	0.98	1.03
7.03	$C_5H_4O_2$	Furan-2-carbaldehyde	n. d.	3.28
8.60	$C_6H_{10}O_2$	3-methylpentane-2,4,-dione	0.28	0.19
9.76	C_6H_8O	2,4-dimethylfuran	0.45	0.13
11.03	$C_5H_6O_2$	2-hydroxycyclopent-2-enone	0.40	0.23
14.06	$C_8H_{12}O_3$	2-hydroxy-6-methycyclohex-3-enecarboxylic acid	n. d.	0.17
15.24	$C_5H_8O_3$	2,3-dihydroxycyclopentanone	0.69	n. d.

续表

时间 (min)	分子式	分子结构	相对含量	
			FSD 生物油	Cu-FSD 生物油
15.34	$C_6H_8O_2$	3-methylcyclopentane-1,2-dione	0.34	0.12
17.41	$C_7H_8O_2$	4-methoxyphenol	5.57	11.23
21.27	$C_8H_{10}O_2$	2-methoxy-4-methylphenol	15.16	12.03
22.29	$C_6H_{12}O_2$	3-methylcyclopentane-1,2-diol	n.d.	0.89
24.31	$C_9H_{12}O_2$	4-ethyl-2-methoxyphenol	3.05	19.69
25.57	$C_9H_{10}O_2$	1-(2-hydroxy-5-methylphenyl) ethanone	5.79	12.95
26.95	$C_8H_{10}O_3$	2,6-dimethoxyphenol	8.69	5.24
27.21	$C_6H_6O_3$	Benzene-1,2,4-triol	0.53	0.82
28.60	$C_{10}H_{12}O_2$	2-methoxy-4-(prop-1-en-1-yl) phenol	19.30	6.20
29.21	$C_{10}H_{10}O_4$	4-formyl-2-methoxyphenyl acatate	0.25	8.75
31.82	$C_{14}H_{22}O$	2,4-di-tert-butylphenol	n.d.	0.95
32.23	$C_{12}H_{19}O_3N$	1-(2,3,5-trimethoxyphenyl)propan-2-amine	1.64	0.92
33.42	$C_{10}H_{12}O_3$	1-(2,4-dimethoxyphenyl) ethanone	6.62	0.11
34.27	$C_{10}H_{14}O_2$	2-hydro-4-methoxy-3,6-dimethyl-bezaldehyde	0.43	n.d.
34.45	$C_{11}H_{14}O_3$	4-allyl-2,6-dimethoxylphenol	10.06	0.18
35.87	$C_{11}H_{14}O_3$	4-ethyl-2,5-dimethoxylbenzaldehyde	0.61	n.d.
36.16	$C_7H_6O_4$	2,4-dihydroxybenzoic acid	n.d.	0.51
39.27	$C_{11}H_{14}O_4$	1-(2,6-dihydroxy-4-methoxyphenyl)butan-1-one	1.46	0.97
40.32	$C_{15}H_{14}O_3$	Benzyl 2-methoxybenzoate	0.51	0.53
43.18	$C_8H_8O_5$	2-(3,4-dihydroxyphenyl)-2-hydroxyacetic acid	0.86	4.17
43.72	$C_{17}H_{18}O_4$	2-(4-(benzyloxy)-3-ethoxyphenyl) acetic acid	2.53	0.50

注：* n.d.表示没有检测到。

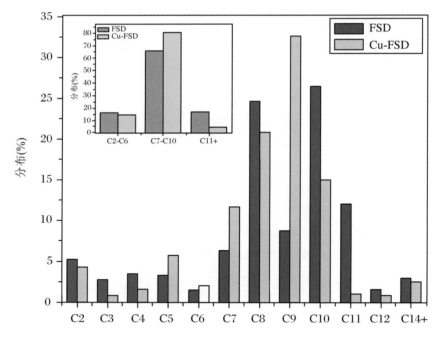

图 3.13 不同碳原子数的化合物在 Cu-FSD 和 FSD 热解生物油中的分布

3.3.4

铜在热解过程中的分布、迁移、转化及回收

铜元素在热解生物炭(Cu(生物炭)%)和生物油(Cu(生物油)%)中的分布可以通过以下公式进行计算:

$$Cu(生物炭)\% = \frac{热解生物炭中的铜含量 \times 热解生物炭的质量}{Cu\text{-}FSD 中的铜含量 \times Cu\text{-}FSD 的质量} \times 100\% \tag{3.4}$$

$$Cu(生物油)\% = \frac{生物油中的铜含量 \times 生物油的质量}{Cu\text{-}FSD 中的铜含量 \times Cu\text{-}FSD 的质量} \times 100\% \tag{3.5}$$

不同温度下热解过程中铜在热解生物炭和生物油中的分布情况如图 3.14(a)所示:Cu(生物油)%随着温度的升高不断上升,而 Cu(生物炭)%则呈现出相反的趋势。由于铜元素的挥发性很差,绝大部分的铜(>90%)在热解后都存在于热解生物炭之中。如表 3.11 所示,铜在热解生物炭中的质量分数在 2.6%~3.9%之间,远远大于其在生物油中的质量分数。在热解生物炭中的铜元素通过一个简单的灼烧就可以加以回收,如图 3.14(b)所示,当含铜的生物炭在 600 ℃

空气中灼烧 4 h 后,所得灰分中铜的含量大于 30%,铜的回收率达到 98.9%,即使灼烧温度达到 900 ℃,铜的回收率也能达到 94.5%。而在生物油中,铜的含量在 4.6~7.6 mg·kg^{-1},这个含量甚至低于国家标准汽油中铜的含量限制,因此,当生物油用于化石燃料替代时,无需担心其中所含的重金属造成二次污染。

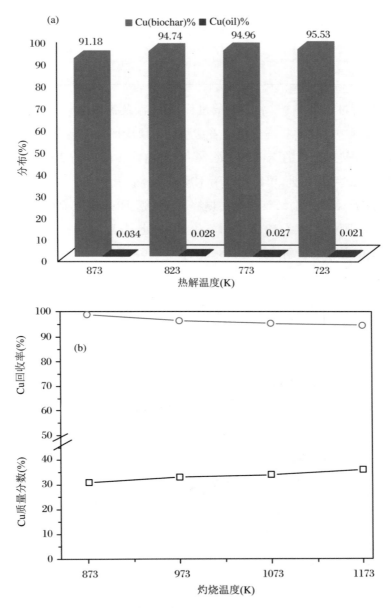

图 3.14 (a) 不同热解温度下铜在热解生物炭和生物油中的分布;
(b) 不同灼烧温度对含铜热解生物炭中铜的回收效率

表 3.11　不同温度下热解所得生物油和热解生物炭中的铜含量

温度 (K)	Cu(生物炭) (wt%)	Cu(生物油) (mg·kg^{-1})
873	3.87	7.58
823	3.74	6.46
773	3.54	5.58
723	2.61	4.61
Cu-FSD	1.09(wt%)	

快速热解过程是一个非常复杂的过程,其中涉及各种化学反应。因此,铜元素在热解过程中的化学状态也将随着这些化学反应而发生转化。图 3.15 所示的是 Cu-FSD 热解前后的 XRD 图谱。在热解前,Cu-FSD 的 XRD 图谱只有一个衍射峰在 $2\theta=22.9°$ 处,可以指认为木质纤维素的衍射峰。在热解之后,这个衍射峰消失了,表明木质纤维结构在热解过程中遭到破坏,取而代之的是 3 个新的衍射峰,分别位于 $2\theta=43.5°,50.7°$ 和 74.2° 处,分别归结为金属铜的(111),(200)和(210)晶面的衍射峰(JCPDS file:04-0836),这表明二价铜经过热解之后转变成了面心立方晶型的零价铜。

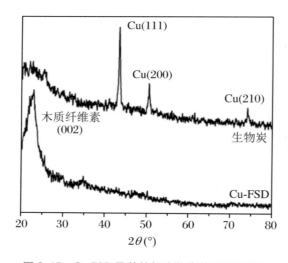

图 3.15　Cu-FSD 及其热解生物炭的 XRD 图谱

热解前后铜的化学状态的变化通过 XPS 进行表征,结果如图 3.16 所示。在二价铜的 XPS 图谱中通常会有一个特征的卫星峰出现在比 Cu 2p 峰高 10 eV 的地方,可以用来区分二价铜和低价态的铜元素[127]。在 Cu-FSD 的 XPS 图谱中,这个卫星峰被观测到在结合能为 943.0 eV 处,比 Cu 2p 3/2 的 XPS 峰高 8.8 eV。在热解之后,卫星峰消失,并且 Cu 2p 3/2 的 XPS 峰的结合能也降低到了 931.2 eV，以上结果都表明在热解过程中，二价铜被还原成了低价态的铜。根据 XPS 和 XRD 的结果，我们推测出铜元素在热解过程中的转化机理（图 3.17），在热解过程中，生物质分子发生裂解，产生大量的小分子还原性物质，例如 H_2、CO、$C_xH_yO_z$，和一些电子（e^-）。而这些电子在热解过程中被具有氧化性的二价铜捕获，形成零价铜。

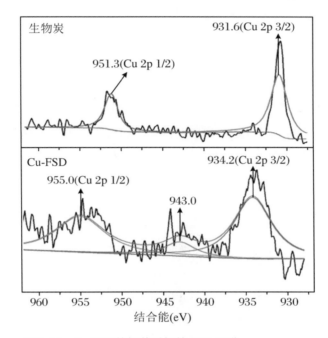

图 3.16 Cu-FSD 热解前后铜的 XPS 图谱

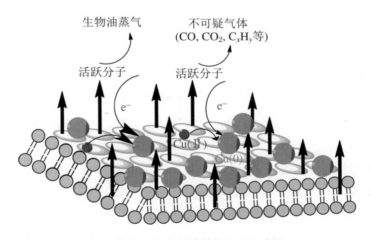

图 3.17 Cu-FSD 热解过程中铜的转化机理示意图

在本节中，我们通过预先向生物质中负载重金属铜，然后再将铜负载生物质进行热解，研究铜元素在热解过程中对生物油品质提升的催化作用及其机理，根

据实验结果,可以得出以下结论:

(1) Cu-FSD 热解产生的生物油的产率和品质,相对于未经负载的生物质热解所得生物油有了明显的提升。

(2) 紫外可见光谱、核磁共振波谱以及 GC-MS 分析结果表明,Cu-FSD 热解产生的生物油中所含的芳香化合物比未负载生物质所产生物油的芳香化合物含量大为增加;铜的存在能促进生物质中木质素组分的分解,从而产生大量的芳香化合物。

(3) 在热解过程中,铜元素主要集中在热解生物炭中,其分布比超过 90%,这部分铜可以通过灼烧的方法加以回收。铜元素在热解过程中从二价铜最终将转变为零价的面心立方晶型的金属铜。考虑到铜负载生物质可以直接使用废弃生物质吸附废水中的重金属铜得到,并且热解所得生物油中的铜元素含量极低,因此该方法可以用作综合处理重金属废水及回收废弃生物质。

参考文献

[1] Van Putten R-J,Van Der Waal J C,De Jong E,et al. Hydroxymethylfurfural,a versatile platform chemical made from renewable resources [J]. Chemical Reviews,2013,113(3):1499-1597.

[2] Takagaki A,Takahashi M,Nishimura S,et al. One-pot synthesis of 2,5-diformylfuran from carbohydrate derivatives by sulfonated resin and hydrotalcite-supported ruthenium catalysts [J]. ACS Catalysis,2011,1(11):1562-1565.

[3] Zhang Z,Liu B,Lv K,et al. Aerobic oxidation of biomass derived 5-hydroxymethylfurfural into 5-hydroxymethyl-2-furancarboxylic acid catalyzed by a montmorillonite K-10 clay immobilized molybdenum acetylacetonate complex [J]. Green Chemistry,2014,16(5):2762-2770.

[4] Lan J,Lin J,Chen Z,et al. Transformation of 5-hydroxymethylfurfural (HMF) to maleic anhydride by aerobic oxidation with heteropolyacid catalysts [J]. ACS Catalysis,2015,5(4):2035-2041.

[5] Gupta N K,Nishimura S,Takagaki A,et al. Hydrotalcite-supported gold-nanoparticle-catalyzed highly efficient base-free aqueous oxidation of 5-hydroxymethylfurfural into 2,5-furandicarboxylic acid under atmospheric oxygen pressure [J]. Green Chemistry,2011,13(4):824-827.

[6] Dijkman W P, Groothuis D E, Fraaije M W. Enzyme-catalyzed oxidation of 5-hydroxymethylfurfural to furan-2,5-dicarboxylic acid [J]. Angewandte Chemie International Edition, 2014, 53(25): 6515-6518.

[7] Jiang N, You B, Boonstra R, et al. Integrating electrocatalytic 5-hydroxymethylfurfural oxidation and hydrogen production via Co-P-derived electrocatalysts [J]. ACS Energy Letters, 2016, 1(2): 386-390.

[8] You B, Liu X, Jiang N, et al. A general strategy for decoupled hydrogen production from water splitting by integrating oxidative biomass valorization [J]. Journal of the American Chemical Society, 2016, 138(41): 13639-13646.

[9] Chadderdon D J, Xin L, Qi J, et al. Electrocatalytic oxidation of 5-hydroxymethylfurfural to 2,5-furandicarboxylic acid on supported Au and Pd bimetallic nanoparticles [J]. Green Chemistry, 2014, 16(8): 3778-3786.

[10] Vuyyuru K R, Strasser P. Oxidation of biomass derived 5-hydroxymethylfurfural using heterogeneous and electrochemical catalysis [J]. Catalysis Today, 2012, 195(1): 144-154.

[11] Rajendran S, Raghunathan R, Hevus I, et al. Programmed photodegradation of polymeric/oligomeric materials derived from renewable bioresources [J]. Angewandte Chemie International Edition, 2015, 54(4): 1159-1163.

[12] Wilsens C H, Wullems N J, Gubbels E, et al. Synthesis, kinetics, and characterization of bio-based thermosets obtained through polymerization of a 2,5-furandicarboxylic acid-based bis(2-oxazoline) with sebacic acid [J]. Polymer Chemistry, 2015, 6(14): 2707-2716.

[13] Wilsens C H R M, Verhoeven J M G A, Noordover B a J, et al. Thermotropic polyesters from 2,5-Furandicarboxylic acid and vanillic acid: Synthesis, thermal properties, melt behavior, and mechanical performance [J]. Macromolecules, 2014, 47(10): 3306-3316.

[14] Zhu J, Cai J, Xie W, et al. Poly(butylene 2,5-furan dicarboxylate), a biobased alternative to PBT: Synthesis, physical properties, and crystal structure [J]. Macromolecules, 2013, 46(3): 796-804.

[15] Bozell J J, Petersen G R. Technology development for the production of biobased products from biorefinery carbohydrates: The US department of energy's "Top 10" revisited [J]. Green Chemistry, 2010, 12(4): 539-554.

[16] Smith P B. Bio-based sources for terephthalic acid [C]// Green Polymer Chemistry: Biobased materials and biocatalysis. American Chemical Society. 2015: 453-469.

[17] Lei D, Yu K, Li M R, et al. Facet effect of single-crystalline Pd nanocrystals for aerobic oxidation of 5-hydroxymethyl-2-furfural [J]. ACS Catalysis, 2017, 7(1): 421-432.

[18] Wan X, Zhou C, Chen J, et al. Base-free aerobic oxidation of 5-hydroxymethyl-furfural to 2,5-furandicarboxylic acid in water catalyzed by functionalized carbon nanotube-supported Au-Pd alloy nanoparticles [J]. ACS Catalysis, 2014, 4(7): 2175-2185.

[19] Wang Y, Yu K, Lei D, et al. Basicity-tuned hydrotalcite-supported Pd catalysts for aerobic oxidation of 5-hydroxymethyl-2-furfural under mild conditions [J]. ACS Sustainable Chemistry & Engineering, 2016, 4(9): 4752-4761.

[20] Wan X, Zhou C, Chen J, et al. Base-free aerobic oxidation of 5-hydroxymethyl-furfural to 2,5-furandicarboxylic acid in water catalyzed by functionalized carbon nanotube-supported Au-Pd alloy nanoparticles [J]. ACS Catalysis, 2014, 4(7): 2175-2185.

[21] Yi G, Teong S P, Zhang Y. Base-free conversion of 5-hydroxymethylfurfural to 2,5-furandicarboxylic acid over a Ru/C catalyst [J]. Green Chemistry, 2016, 18(4): 979-983.

[22] Cha H G, Choi K-S. Combined biomass valorization and hydrogen production in a photoelectrochemical cell [J]. Nat Chem, 2015, 7(4): 328-333.

[23] You B, Jiang N, Liu X, et al. Simultaneous H_2 generation and biomass upgrading in water by an efficient noble-metal-free bifunctional electrocatalyst [J]. Angewandte Chemie International Edition, 2016, 55(34): 9913-9917.

[24] You B, Liu X, Liu X, et al. Efficient H_2 evolution coupled with oxidative refining of alcohols via a hierarchically porous nickel bifunctional electrocatalyst [J]. ACS Catalysis, 2017.

[25] Stern L-A, Feng L, Song F, et al. Ni_2P as a Janus catalyst for water splitting: The oxygen evolution activity of Ni_2P nanoparticles [J]. Energy & Environmental Science, 2015, 8(8): 2347-2351.

[26] Dutta A, Samantara A K, Dutta S K, et al. Surface-oxidized dicobalt phosphide nanoneedles as a nonprecious, durable, and efficient OER catalyst [J]. ACS Energy Letters, 2016, 1(1): 169-174.

[27] Jin S. Are metal chalcogenides, nitrides, and phosphides oxygen evolution catalysts or bifunctional catalysts? [J]. ACS Energy Letters, 2017: 1937-1938

[28] Mccrory C C L, Jung S, Peters J C, et al. Benchmarking heterogeneous electrocatalysts for the oxygen evolution reaction [J]. Journal of the American Chemical Society, 2013, 135(45): 16977-16987.

[29] Hunter B M, Gray H B, Müller A M. Earth-abundant heterogeneous water oxidation catalysts [J]. Chemical Reviews, 2016, 116(22): 14120-14136.

[30] Duan X, Evans D G. Layered double hydroxides [M]. Berlin; New York: Springer, 2005.

[31] Liang H, Meng F, Cabán-Acevedo M, et al. Hydrothermal continuous flow synthesis and exfoliation of NiCo layered double hydroxide nanosheets for enhanced oxygen evolution catalysis [J]. Nano Letters, 2015, 15(2): 1421-1427.

[32] Zhu X, Tang C, Wang H-F, et al. Monolithic-structured ternary hydroxides as freestanding bifunctional electrocatalysts for overall water splitting [J]. J Mater Chem A, 2016, 4(19): 7245-7250.

[33] Song F, Hu X. Exfoliation of layered double hydroxides for enhanced oxygen evolution catalysis [J]. Nat Commun, 2014, 5: 4477.

[34] Zhang B, Zheng X, Voznyy O, et al. Homogeneously dispersed, multimetal oxygen-evolving catalysts [J]. Science, 2016, 352(6283): 333-337.

[35] Lu Z, Qian L, Tian Y, et al. Ternary NiFeMn layered double hydroxides as highly-efficient oxygen evolution catalysts[J]. Chemical Communications, 2016, 52(5): 908-911.

[36] Chen J Y C, Dang L, Liang H, et al. Operando analysis of NiFe and Fe oxyhydroxide electrocatalysts for water oxidation: Detection of Fe^{4+} by mössbauer spectroscopy [J]. Journal of the American Chemical Society, 2015, 137(48): 15090-15093.

[37] Gul S, Ng J W D, Alonso-Mori R, et al. Simultaneous detection of electronic structure changes from two elements of a bifunctional catalyst using wavelength-dispersive X-ray emission spectroscopy and in situ

[38] Dai L, Qin Q, Zhao X, et al. Electrochemical partial reforming of ethanol into ethyl acetate using ultrathin Co_3O_4 nanosheets as a highly selective anode catalyst [J]. ACS Central Science, 2016, 2(8): 538-544.

[39] Forticaux A, Dang L, Liang H, et al. Controlled synthesis of layered double hydroxide nanoplates driven by screw dislocations [J]. Nano Letters, 2015, 15(5): 3403-3409.

[40] Okamoto K, Iyi N, Sasaki T. Factors affecting the crystal size of the MgAl-LDH (layered double hydroxide) prepared by using ammonia-releasing reagents [J]. Applied Clay Science, 2007, 37(1): 23-31.

[41] De Waal S A, Viljoen E A. Nickel minerals from barberton, south africa. IV: Reevesite, a member of the hydrotalcite group [J]. American Mineralogist, 1971, 56(5): 1077-1081.

[42] Louie M W, Bell A T. An investigation of thin-film Ni-Fe oxide catalysts for the electrochemical evolution of oxygen [J]. Journal of the American Chemical Society, 2013, 135(33): 12329-12337.

[43] Zboril R, Mashlan M, Petridis D. Iron (III) oxides from thermal processes synthesis, structural and magnetic properties, Mössbauer spectroscopy characterization, and applications [J]. Chemistry of Materials, 2002, 14(3): 969-982.

[44] Yu L, Zhou H, Sun J, et al. Cu nanowires shelled with NiFe layered double hydroxide nanosheets as bifunctional electrocatalysts for overall water splitting [J]. Energy & Environmental Science, 2017, 10: 1820-1827.

[45] Gong M, Li Y, Wang H, et al. An advanced Ni-Fe layered double hydroxide electrocatalyst for water oxidation [J]. Journal of the American Chemical Society, 2013, 135(23): 8452-8455.

[46] Zuo X, Venkitasubramanian P, Busch D H, et al. Optimization of Co/Mn/Br-catalyzed oxidation of 5-hydroxymethylfurfural to enhance 2,5-furandicarboxylic acid yield and minimize substrate burning [J]. ACS Sustainable Chemistry & Engineering, 2016, 4(7): 3659-3668.

[47] Huber G W, Iborra S, Corma A. Synthesis of transportation fuels from biomass: Chemistry, catalysts, and engineering [J]. Chemical Reviews, 2006, 106(9): 4044-4098.

electrochemistry [J]. Physical Chemistry Chemical Physics, 2015, 17(14): 8901-8912.

[48] Czernik S, Bridgwater A V. Overview of applications of biomass fast pyrolysis oil [J]. Energy & Fuels, 2004, 18(2): 590-598.

[49] Kunkes E L, et al. Catalytic conversion of biomass to monofunctional hydrocarbons and targeted liquid-fuel classes [J]. Science, 2008, 322: 417.

[50] Rahman M M, Liu R, Cai J. Catalytic fast pyrolysis of biomass over zeolites for high quality bio-oil: A review [J]. Fuel Processing Technology, 2018, 180: 32-46.

[51] Ni S, Liu R, Rahman M M, et al. A review on the catalytic pyrolysis of biomass for the bio-oil production with ZSM-5: Focus on structure [J]. Fuel Processing Technology, 2020, 199: 106301.

[52] Mutsengerere S, Chihobo C H, Musademba D, et al. A review of operating parameters affecting bio-oil yield in microwave pyrolysis of lignocellulosic biomass [J]. Renewable and Sustainable Energy Reviews, 2019, 104: 328-336.

[53] Liu R, Sarker M, Rahman M M, et al. Multi-scale complexities of solid acid catalysts in the catalytic fast pyrolysis of biomass for bio-oil production: A review [J]. Progress in Energy and Combustion Science, 2020, 80: 100852.

[54] Li F, Srivatsa S C, Bhattacharya S. A review on catalytic pyrolysis of microalgae to high-quality bio-oil with low oxygeneous and nitrogenous compounds [J]. Renewable and Sustainable Energy Reviews, 2019, 108: 481-497.

[55] Leng L J, Li H, Yuan X Z, et al. Bio-oil upgrading by emulsification/microemulsification: A review [J]. Energy, 2018, 161: 214-232.

[56] Kumar R, Strezov V, Weldekidan H, et al. Lignocellulose biomass pyrolysis for bio-oil production: A review of biomass pre-treatment methods for production of drop-in fuels [J]. Renewable and Sustainable Energy Reviews, 2020, 123: 109763.

[57] Hansen S, Mirkouei A, Diaz L A. A comprehensive state-of-technology review for upgrading bio-oil to renewable or blended hydrocarbon fuels [J]. Renewable and Sustainable Energy Reviews, 2020, 118: 109548.

[58] Alvarez-Chavez B J, Godbout S, Palacios-Rios J H, et al. Physical, chemical, thermal and biological pre-treatment technologies in fast pyrolysis to maximize bio-oil quality: A critical review [J]. Biomass and Bioenergy, 2019, 128: 105333.

[59] Mohan D, Pittman C U, Steele P H. Pyrolysis of wood/biomass for bio-oil: A critical review [J]. Energy & Fuels, 2006, 20(3): 848-889.

[60] Tang Z, Lu Q, Zhang Y, et al. One-step bio-oil upgrading through hydrotreatment, esterification, and cracking [J]. Industrial & Engineering Chemistry Research, 2009, 48(15): 6923-6929.

[61] Sipilä K, Kuoppala E, Fagernäs L, et al. Characterization of biomass-based flash pyrolysis oils [J]. Biomass and Bioenergy, 1998, 14(2): 103-113.

[62] Friedl A, Padouvas E, Rotter H, et al. Prediction of heating values of biomass fuel from elemental composition [J]. Analytica Chimica Acta, 2005, 544(1/2): 191-8.

[63] Edward F. Catalytic hydrodeoxygenation [J]. Applied Catalysis A: General, 2000, 199(2): 147-190.

[64] Li N, Tompsett G A, Huber G W. Renewable high-octane gasoline by aqueous-phase hydrodeoxygenation of C5 and C6 carbohydrates over Pt/Zirconium phosphate catalysts [J]. ChemSusChem, 2010, 3(10): 1154-1157.

[65] Vispute T P, Huber G W. Production of hydrogen, alkanes and polyols by aqueous phase processing of wood-derived pyrolysis oils [J]. Green Chemistry, 2009, 11(9): 1433.

[66] De Miguel Mercader F, Groeneveld M J, Kersten S R A, et al. Hydrodeoxygenation of pyrolysis oil fractions: Process understanding and quality assessment through co-processing in refinery units [J]. Energy & Environmental Science, 2011, 4(3): 985.

[67] Zhao H Y, Li D, Bui P, et al. Hydrodeoxygenation of guaiacol as model compound for pyrolysis oil on transition metal phosphide hydroprocessing catalysts [J]. Applied Catalysis A: General, 2011, 391(1/2): 305-310.

[68] Wang W, Yang Y, Luo H, et al. Amorphous Co-Mo-B catalyst with high activity for the hydrodeoxygenation of bio-oil [J]. Catalysis Communications, 2011, 12(6): 436-440.

[69] Rao T V M, Clavero M M, Makkee M. Effective gasoline production strategies by catalytic cracking of rapeseed vegetable oil in refinery conditions [J]. ChemSusChem, 2010, 3(7): 807-810.

[70] Zhang Z J, Wang Q W, Tripathi P, et al. Catalytic upgrading of bio-oil using 1-octene and 1-butanol over sulfonic acid resin catalysts [J]. Green

Chemistry, 2011, 13(4): 940.

[71] Rioche C, Kulkarni S, Meunier F C, et al. Steam reforming of model compounds and fast pyrolysis bio-oil on supported noble metal catalysts [J]. Applied Catalysis B: Environmental, 2005, 61(1/2): 130-139.

[72] Czernik S, French R, Feik C, et al. Hydrogen by catalytic steam reforming of liquid byproducts from biomass thermoconversion processes [J]. Industrial & Engineering Chemistry Research, 2002, 41(17): 4209-4215.

[73] Yu W, Tang Y, Mo L, et al. One-step hydrogenation-esterification of furfural and acetic acid over bifunctional Pd catalysts for bio-oil upgrading [J]. Bioresource Technology, 2011, 102(17): 8241-8246.

[74] Tang Y, Yu W, Mo L, et al. One-step hydrogenation-esterification of aldehyde and acid to ester over bifunctional Pt catalysts: A model reaction as novel route for catalytic upgrading of fast pyrolysis bio-oil [J]. Energy & Fuels, 2008, 22(5): 3484-3488.

[75] Mahfud F H, Ghijsen F, Heeres H J. Hydrogenation of fast pyrolyis oil and model compounds in a two-phase aqueous organic system using homogeneous ruthenium catalysts [J]. Journal of Molecular Catalysis A: Chemical, 2007, 264(1/2): 227-236.

[76] Reuter R, Wegner H A. Synthesis and isomerization studies of cyclotrisazobiphenyl [J]. Chemistry, 2011, 17(10): 2987-2995.

[77] Parks D J, Parsons W H, Colburn R W, et al. Design and optimization of benzimidazole-containing Transient Receptor Potential Melastatin 8 (TRPM8) antagonists [J]. Journal of Medicinal Chemistry, 2010, 54(1): 233-247.

[78] Marvania B, Lee P-C, Chaniyara R, et al. Design, synthesis and antitumor evaluation of phenyl N-mustard-quinazoline conjugates [J]. Bioorganic & Medicinal Chemistry, 2011, 19(6): 1987-1998.

[79] Chanda K, Maiti B, Chung W-S, et al. Novel approach towards 2-substituted aminobenzimidazoles on imidazolium ion tag under focused microwave irradiation [J]. Tetrahedron, 2011, 67(34): 6214-6220.

[80] Trost B M, O'boyle B M, Torres W, et al. Development of a flexible strategy towards FR900482 and the mitomycins [J]. Chemistry, 2011, 17(28): 7890-7903.

[81] Hu M, Li L, Wu H, et al. Multicolor, one- and two-photon imaging of

enzymatic activities in live cells with fluorescently quenched Activity-Based Probes (qABPs) [J]. Journal of the American Chemical Society, 2011, 133 (31): 12009-12020.

[82] Clemmensen E. Chemische Berichte[M]. 1913, 46: 1837.

[83] Di Vona M L, Floris B, Luchetti L, et al. Single electron transfers in zinc-promoted reactions. The mechanisms of the clemmensen reduction and related reactions [J]. Tetrahedron Letters, 1990, 31(42): 6081-6084.

[84] Xu S, Toyama T, Nakamura J, et al. One-pot reductive cleavage of exo-olefin to methylene with a mild ozonolysis-Clemmensen reduction sequence [J]. Tetrahedron Letters, 2010, 51(34): 4534-4537.

[85] Tee Y-H, Grulke E, Bhattacharyya D. Role of Ni/Fe nanoparticle composition on the degradation of trichloroethylene from water [J]. Industrial & Engineering Chemistry Research, 2005, 44(18): 7062-7070.

[86] Peralta M A, Sooknoi T, Danuthai T, et al. Deoxygenation of benzaldehyde over CsNaX zeolites [J]. Journal of Molecular Catalysis A: Chemical, 2009, 312(1/2): 78-86.

[87] Ausavasukhi A, Sooknoi T, Resasco D E. Catalytic deoxygenation of benzaldehyde over gallium-modified ZSM-5 zeolite [J]. Journal of Catalysis, 2009, 268(1): 68-78.

[88] Okamoto M, Hirao T, Yamaai T. Polymers as novel modifiers for supported metal catalyst in hydrogenation of benzaldehydes [J]. Journal of Catalysis, 2010, 276(2): 423-428.

[89] Saadi A, Merabti R, Rassoul Z, et al. Benzaldehyde hydrogenation over supported nickel catalysts [J]. Journal of Molecular Catalysis A: Chemical, 2006, 253(1/2): 79-85.

[90] Radhakrishan R, Do D M, Jaenicke S, et al. Potassium phosphate as a solid base catalyst for the catalytic transfer hydrogenation of aldehydes and ketones [J]. ACS Catalysis, 2011, 1(11): 1631-1636.

[91] Smith M B, March J. March's advanced organic chemistry[M]. 5th ed. New York: John Wiley & Sons, 2001.

[92] Scholze B, Meier D. Characterization of the water-insoluble fraction from pyrolysis oil (pyrolytic lignin). Part I: PY-GC/MS, FTIR, and functional groups [J]. Journal of Analytical and Applied Pyrolysis, 2001, 60(1): 41-54.

[93] Fahmi R, Bridgwater A V, Donnison I, et al. The effect of lignin and inorganic species in biomass on pyrolysis oil yields, quality and stability [J]. Fuel, 2008, 87(7): 1230-1240.

[94] Kosa M, Ben H, Theliander H, et al. Pyrolysis oils from CO_2 precipitated kraft lignin [J]. Green Chemistry, 2011, 13(11): 3196.

[95] Desisto W J, Hill N, Beis S H, et al. Fast pyrolysis of pine sawdust in a fluidized-bed reactor [J]. Energy & Fuels, 2010, 24(4): 2642-2651.

[96] Mullen C A, Strahan G D, Boateng A A. Characterization of various fast-pyrolysis bio-oils by NMR spectroscopy [J]. Energy & Fuels, 2009, 23(5): 2707-2718.

[97] Zakzeski J, Bruijnincx P C A, Jongerius A L, et al. The catalytic valorization of lignin for the production of renewable chemicals [J]. Chemical Reviews, 2010, 110(6): 3552-3599.

[98] Patwardhan P R, Brown R C, Shanks B H. Understanding the fast pyrolysis of lignin [J]. ChemSusChem, 2011, 4(11): 1629-1636.

[99] Calvo-Flores F G, Dobado J A. Lignin as renewable raw material [J]. ChemSusChem, 2010, 3(11): 1227-1235.

[100] Petrus L, Noordermeer M A. Biomass to biofuels, a chemical perspective [J]. Green Chemistry, 2006, 8(10): 861.

[101] Sergeev A G, Hartwig J F. Selective, nickel-catalyzed hydrogenolysis of aryl ethers [J]. Science, 2011, 332(6028): 439-443.

[102] Rydén M, Johansson M, Lyngfelt A, et al. NiO supported on Mg-ZrO_2 as oxygen carrier for chemical-looping combustion and chemical-looping reforming [J]. Energy & Environmental Science, 2009, 2(9): 970-981.

[103] Matsumura Y, Ishibe H. High temperature steam reforming of methanol over Cu/ZnO/ZrO_2 catalysts [J]. Applied Catalysis B: Environmental, 2009, 91(1/2): 524-532.

[104] Lee Y, Hoveyda A H. Efficient boron-copper additions to aryl-substituted alkenes promoted by NHC-based catalysts. Enantioselective Cu-catalyzed hydroboration reactions [J]. Journal of the American Chemical Society, 2009, 131(9): 3160-3161.

[105] Roy P, Periasamy A P, Liang C-T, et al. Synthesis of graphene-ZnO-Au nanocomposites for efficient photocatalytic reduction of nitrobenzene [J].

Environmental Science & Technology, 2013, 47(12): 6688-6695.

[106] Xu F, Deng S, Xu J, et al. Highly active and stable Ni-Fe bimetal prepared by ball milling for catalytic hydrodechlorination of 4-Chlorophenol [J]. Environmental Science & Technology, 2012, 46(8): 4576-4582.

[107] Janas J, Gurgul J, Socha R P, et al. Effect of Cu content on the catalytic activity of CuSiBEA zeolite in the SCR of NO by ethanol: Nature of the copper species [J]. Applied Catalysis B: Environmental, 2009, 91(1/2): 217-224.

[108] Lin Y C, Huber G W. The critical role of heterogeneous catalysis in lignocellulosic biomass conversion [J]. Energy & Environmental Science, 2009, 2(1): 68-80.

[109] Zheng M Y, Wang A Q, Ji N, et al. Transition metal-tungsten bimetallic catalysts for the conversion of cellulose into ethylene glycol [J]. ChemSusChem, 2010, 3(1): 63-66.

[110] Kunkes E L, Simonetti D A, West R M, et al. Catalytic conversion of biomass to monofunctional hydrocarbons and targeted liquid-fuel classes [J]. Science, 2008, 322(5900): 417-421.

[111] Fan Y, Xiao H, Shi G, et al. A novel approach for modulating the morphology of supported metal nanoparticles in hydrodesulfurization catalysts [J]. Energy & Environmental Science, 2011, 4(2): 572.

[112] Landong L, Jixin C, Shujuan Z, et al. Selective catalytic reduction of nitrogen oxides from exhaust of lean burn engine over in-situ synthesized Cu-ZSM-5/cordierite [J]. Environmental Science & Technology, 2005, 39(8): 2841-2847.

[113] Chang S H, Yeh J W, Chein H M, et al. PCDD/F adsorption and destruction in the flue gas streams of MWI and MSP via Cu and Fe catalysts supported on carbon [J]. Environmental Science & Technology, 2008, 42(15): 5727-5733.

[114] Saranya K, Sundaramanickam A, Shekhar S, et al. Biosorption of multi-heavy metals by coral associated phosphate solubilising bacteria Cronobacter muytjensii KSCAS2 [J]. Journal of Environmental Management, 2018, 222: 396-401.

[115] Sahmoune M N. Performance of streptomyces rimosus biomass in

biosorption of heavy metals from aqueous solutions [J]. Microchemical Journal, 2018, 141: 87-95.

[116] Najafpour A, Rajabi Khorrami A, Aberoomand Azar P, et al. Study of heavy metals biosorption by tea fungus in Kombucha drink using Central Composite Design [J]. Journal of Food Composition and Analysis, 2020, 86: 103359.

[117] Jin Z, Deng S, Wen Y, et al. Application of simplicillium chinense for Cd and Pb biosorption and enhancing heavy metal phytoremediation of soils [J]. Science of the Total Environment, 2019, 697: 134148.

[118] Beni A A, Esmaeili A. Biosorption, an efficient method for removing heavy metals from industrial effluents: A review [J]. Environmental Technology & Innovation, 2020, 17: 100503.

[119] Skutil K, Czechowicz D, Taniewski M. Nitrogen-rich natural gases as a potential direct feedstock for some novel methane transformation processes. Part 2: Non-oxidative processes [J]. Energy & Fuels, 2009, 23(9): 4449-4459.

[120] Zhang H, Xiao R, Wang D, et al. Catalytic fast pyrolysis of biomass in a fluidized bed with fresh and spent fluidized catalytic cracking (FCC) catalysts [J]. Energy & Fuels, 2009, 23(12): 6199-6206.

[121] Singh S K, Srinivasa Reddy M, Mangle M, et al. Cu(I)-mediated deoxygenation of N-oxides to amines [J]. Tetrahedron, 2007, 63(1): 126-130.

[122] Cho J, Davis J M, Huber G W. The intrinsic kinetics and heats of reactions for cellulose pyrolysis and char formation [J]. ChemSusChem, 2010, 3(10): 1162-1165.

[123] Wang P, Zhan S, Yu H, et al. The effects of temperature and catalysts on the pyrolysis of industrial wastes (herb residue) [J]. Bioresource Technology, 2010, 101(9): 3236-3241.

[124] Amen-Chen C, Pakdel H, Roy C. Production of monomeric phenols by thermochemical conversion of biomass: A review [J]. Bioresource Technology, 2001, 79(3): 277-299.

[125] Tsai W, Lee M, Chang Y. Fast pyrolysis of rice husk: Product yields and compositions [J]. Bioresource Technology, 2007, 98(1): 22-28.

[126] Azeez A M, Meier D, Odermatt J R, et al. Fast pyrolysis of african and european lignocellulosic biomasses using Py-GC/MS and fluidized bed reactor [J]. Energy & Fuels, 2010, 24(3): 2078-2085.

[127] Arena F, Barbera K, Italiano G, et al. Synthesis, characterization and activity pattern of Cu-ZnO/ZrO$_2$ catalysts in the hydrogenation of carbon dioxide to methanol [J]. Journal of Catalysis, 2007, 249(2): 185-194.

第 4 章

生物炭基功能材料的设计合成

4.1
$MgCl_2$ 负载生物质热解制备介孔碳支撑的纳米氧化镁材料及其对 CO_2 捕集性能研究

4.1.1
概述

大气中不断增加的 CO_2 的浓度对全球气候平衡造成了严重的威胁[1]。全球从化石燃料燃烧和其他工业活动中释放的 CO_2 的量以每年 3% 的速度递增[2]，在 2010 年达到 330 亿吨[3]。CO_2 的不断释放将地球置于不可逆的气候变暖的趋势中，对全球的生态系统造成严重的威胁[4]。为了缓解全球变暖的趋势，CO_2 减排是一个不得不考虑的问题。CO_2 捕集是减少其排放的一个主要手段。目前常用的 CO_2 捕集方法是使用固体可再生的吸附剂对 CO_2 进行吸附固定[5]，常用的 CO_2 吸附剂包括功能碳材料[6]、氨基固体材料[7]和碱土金属氧化物及其复合材料等[5]。其中，碱土金属氧化物，主要是氧化钙（CaO）基材料，是使用最为广泛的 CO_2 捕集材料[8-16]。氧化钙基材料对 CO_2 的捕集主要是通过 CaO 循环过程来实现的（图 4.1(a)）：CaO 和 CO_2 在 600 ℃反应生成 $CaCO_3$，达到将 CO_2 固定的效果，生成的 $CaCO_3$ 在 800 ℃分解再生成 CaO，形成一个封闭的循环，达到将 CO_2 捕集的目的[17]。然而，由于 $CaCO_3$ 的高度热稳定性，从 $CaCO_3$ 中再生 CaO 通常需要高的能源消耗，可能造成新的 CO_2 释放。此外，高温再生过程难以避免材料的烧结，将大大降低再生材料的 CO_2 捕集性能。

和氧化钙基材料相比，MgO 可以在温度低于 200 ℃的条件下捕集 CO_2，并且其再生温度也很低（300～500 ℃），可以很好地避免烧结问题，也能大大降低再生过程中的能源消耗[18]。然而，购买的 MgO 的 CO_2 捕集能力通常很低（～0.5 mol·kg^{-1}）[19]。MgO 对 CO_2 的捕集位点主要是 O^{2-}—Mg^{2+} 键中的碱性 O^{2-} 离子，而 O^{2-} 离子的碱性强度取决于它的配位数。通常位于晶面棱角和边缘的 O^{2-} 离子比那些位于面上的离子具有更高的碱性，对 CO_2 具有更强的捕集能力[20]。因此,开发出具有高碱性 O^{2-} 离子的 MgO 材料对于提高其吸附性能是十分必要的。

氧化镁颗粒由于尺寸极小,晶面中具有更多的棱角和边缘,可以拥有更多的高碱性 O^{2-} 离子。因此,和大体积的 MgO 材料相比,它拥有更多的活性位点用于 CO_2 的捕集[21]。此外,值得注意的是,由于纳米 MgO 的高碱性,其对 CO_2 的捕集温度也将大大降低,甚至低于 100 ℃[5]。为了避免纳米颗粒的团聚,通常需要一个多孔的载体来稳定它们,而其中碳材料是使用最为广泛的载体材料,这主要是由于其多孔和高比表面积的特性[22]。在本节研究中,我们采用一个简单可持续的方法制备介孔碳稳定的 MgO 纳米材料,该方法通过热解 $MgCl_2$ 负载的生物质可以直接获得所需材料。$MgCl_2$ 在海水中大量的存在(海水中 $MgCl_2$ 的平均浓度为 0.45%)[23],而 $MgCl_2$ 负载的生物质可以通过使用废弃生物质从海

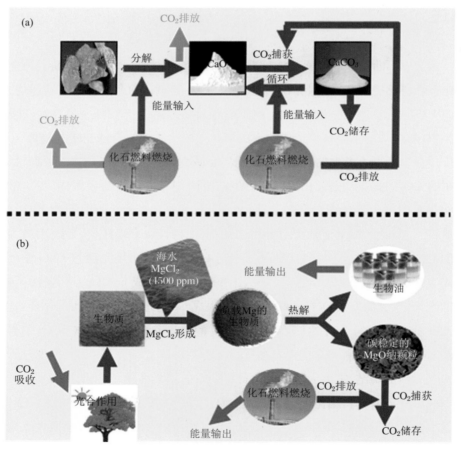

图 4.1 (a) 常规的 CaO 循环方法用于 CO_2 捕集过程的示意图;(b) 介孔碳稳定的氧化镁颗粒的合成及其用于 CO_2 捕集过程示意图;(c) 热解过程中生物质分解过程及介孔碳和氧化镁形成过程的机理示意图

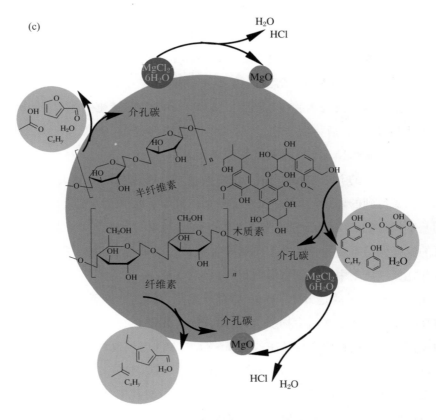

续图 4.1 (a) 常规的 CaO 循环方法用于 CO_2 捕集过程的示意图;(b)介孔碳稳定的氧化镁颗粒的合成及其用于 CO_2 捕集过程示意图;(c) 热解过程中生物质分解过程及介孔碳和氧化镁形成过程的机理示意图

水中吸附 $MgCl_2$ 而方便获得,因此该方法可能得到大规模的应用。在热解过程中,$MgCl_2$ 水解并且分解形成 MgO 纳米颗粒,而生物质分解碳化形成介孔碳材料作为载体用于稳定形成的 MgO 纳米颗粒。热解过程中除了生成固相的介孔碳稳定的 MgO 纳米材料,生物质热解产生的挥发组分经过冷凝液可以获得生物油,可以用于化石燃料替代或者提取化学品。在上述过程中,不需要额外的能源消耗,保证了该过程的可持续性。

4.1.2

介孔碳稳定的氧化镁纳米材料的合成

本次实验所用生物质为杉木木屑,是从合肥本地一个木材处理厂收集而来的,使用之前,将木屑先通过一个高速粉碎机进行粉碎,过筛收集粒径小于 120 目的颗粒,然后干燥 24 h 去除其中的水分。$MgCl_2$ 负载的生物质通过使用木屑生物质作为吸附剂从模拟海水中吸附 $MgCl_2$ 而得,其过程如下:将 5.0 g 的木屑生物质和 1000 mL 的 $MgCl_2$ 溶液混合(其中的镁浓度为 1540 mg·L^{-1},和海水中的平均镁浓度一致),然后在常温下以 200 r·min^{-1} 的转速搅拌 300 min 使吸附过程达到平衡,之后将混合物过滤,所得的固体部分在 105 ℃ 下干燥到恒重,然后再过筛收集粒径小于 120 目的颗粒备用。

介孔碳稳定的 MgO 纳米材料(mPC-MgO)是通过热解 $MgCl_2$ 负载的木屑生物质制备得到的,热解过程是在一个垂直下落固定床反应器中进行的,热解过程的基本操作和 2.2.4 小节的描述基本一致。不同的是热解结束后,反应器加热区的温度保持不变并持续 3 h 使热解生物炭进一步碳化,最后反应器移出加热区并在氮气氛围中冷却到室温获得 mPC-MgO。

图 4.1(b)所示的是 mPC-MgO 的合成路径:首先,水溶液中的 $MgCl_2$ 通过生物吸附作用负载于木屑生物质上,然后 $MgCl_2$ 负载的生物质经历一个快速热解的过程。热解过程中温度从室温在 1~2 s 的时间内迅速升高至 500 ℃ 的热解温度,在如此高的温度下,生物质的组成部分(纤维素、木质素、半纤维素等)快速降解成挥发性物质,冷凝形成生物油(图 4.1(c)),而其中负载的 $MgCl_2$ 水合物经历一系列的脱水和分解反应[24],形成 MgO 纳米颗粒。

$$MgCl_2 \cdot 6H_2O \longrightarrow MgCl_2 \cdot 2H_2O + 4H_2O \uparrow \quad (4.1)$$

$$MgCl_2 \cdot 2H_2O \longrightarrow Mg(OH)Cl + H_2O \uparrow + HCl \uparrow \quad (4.2)$$

$$Mg(OH)Cl \longrightarrow MgO + HCl \uparrow \quad (4.3)$$

尽管在实际应用中,海水中除了镁离子还有其他大量存在的离子,例如 Na^+,K^+,Ca^{2+} 等,但是这些离子由于亲水性极强,很不容易被生物质吸附,而且它们的化合物通常十分稳定,即使在热解温度达到 500 ℃ 时,也不会产生明显的分解,因此对氧化镁纳米颗粒的形成不会产生影响。

4.1.3

介孔碳稳定的氧化镁纳米材料结构和组成表征

mPC-MgO 的主要元素(C,H,N,O)组成通过元素分析仪(VARIO EL III,Elementar Inc.,Germany)进行测定,其中的镁元素要先通过浓硫酸、过氧化氢消解后使用原子发射光谱进行测定。mPC-MgO 的 XRD 分析在一个

18 kW 的旋转阴极 X 射线衍射仪上进行。样品的扫描角度为 20°～80°,扫描速率为 0.02°·s^{-1},所得的衍射峰的归类是通过和 JCPDS 标准卡片对比后确定的。mPC-MgO 的热稳定性通过一个 DTG-60H/DSC-60 热重分析仪,在氮气氛围下以 5 ℃·min^{-1} 从室温升高至 500 ℃来表征。mPC-MgO 的结构特征通过氮气吸脱附等温线来分析,在 -196 ℃下使用物理吸附装置(ASAP 2020 M+C,Micromeritics Co. USA)进行。其比表面积通过 BET 方法进行计算,而孔体积则是通过在相对压力为 0.99 时氮气的吸附量进行计算。mPC-MgO 的表面形貌通过扫描电子显微镜和透射电子显微镜进行观测,其微区组成通过耦合的 X 射线能谱进行测定。

表 4.1 所示的是不同温度下合成的 mPC-MgO 的元素组成和结构特征(比表面积、孔容积和平均孔径等)。从中可知,随着热解温度的升高,mPC-MgO 的镁元素含量不断下降,表明在热解过程中,有相当一部分的镁元素挥发出去了,这和其他金属化合物($CuCl_2$ 和 $PbCl_2$)是不同的[25-26]。这种现象主要可以解释如下:在热解的条件下(高的升温速率和热解温度),一部分的水合氯化镁($MgCl_2·6H_2O$)可能会经历一个剧烈的脱水过程,直接形成无水氯化镁[24],在高的热解温度下直接挥发而离开热解生物炭相中,从而导致 mPC-MgO 中的镁元素含量不断下降,而和 $MgCl_2$ 相比,$CuCl_2$ 和 $PbCl_2$ 具有更强的水解趋势,在热解过程中,不可能直接脱水生成相应的氯化物,而是水解成氢氧化物,随后分解成高沸点的氧化物而留在热解生物炭中。

表 4.1 mPC-MgO 的元素组成和结构特征分析

mPC-MgO	比表面积 ($m^2·g^{-1}$)	孔容积 ($cm^3·g^{-1}$)	平均孔径 (nm)	MgO 晶体尺寸(nm)①	元素组成(wt%)			
					C	H	O②	Mg
mPC-MgO-773	279	0.140	2.01	17.1	56.3	1.1	21.5	21.1
mPC-MgO-873	306	0.156	2.03	18.6	56.7	1.1	21.7	20.5
mPC-MgO-973	298	0.152	2.04	17.8	57.0	1.6	22.0	19.4
mPC-773	14.0	0.067	33.2	—	76.7	3.5	19.8	—

注:① 根据 XRD 图谱中氧化镁(200)衍射峰,通过 Scherrer 方程计算得到。
② 通过差值计算:$O\% = 100\% - C\% - H\% - Mg\%$。

图 4.2(a)所示的是不同热解温度下制备的 mPC-MgO 的 XRD 图谱。这些图谱中都出现了 5 个衍射峰,分别位于衍射角度为 37.1°、43.1°、62.5°、74.7°和 78.6°位置,可以分别归因为氧化镁晶体的(111),(200),(220),(311)和(222)晶面,证实了在热解过程中氧化镁晶体的形成。热解温度从 500 ℃升高至 700 ℃

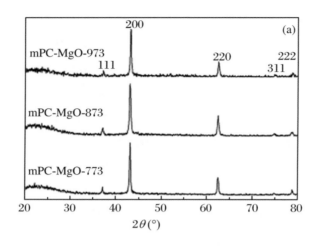

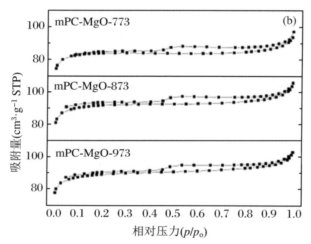

图 4.2 (a) 不同热解温度下制备的 mPC-MgO 的 XRD 图谱；(b) 不同热解温度下制备的 mPC-MgO 的氮气吸脱附等温线

时，氧化镁晶体衍射峰的强度和位置都没有变化，表明氧化镁晶体的热稳定性很高。氧化镁晶体的尺寸可以通过 Scherrer 公式进行计算：

$$d = \frac{K\lambda}{B\cos\theta} \quad (4.4)$$

其中，d(nm)是晶体的尺寸，K 是一个依赖于衍射目标几何形状的一个常数（对于氧化镁，$K=0.94$），λ 是 X 射线的波长（$\lambda=0.154056$ nm），θ(rad)是入射和衍射粒子束之间的角度，B(rad)是半峰宽。根据计算结果，氧化镁的晶体尺寸分别为 17.1 nm，18.6 nm 和 17.8 nm。

mPC-MgO 的结构特征通过氮气吸/脱附方法来表征，结果如图 4.2(b)所示,不同热解温度下获得的 mPC-MgO 氮气吸/脱附等温线都呈现出第Ⅳ模式[27],表明了介孔的存在,其滞回线的模式为在较高相对压力下的 H3 模式,表明了不规则形状和尺寸的带状孔隙的存在[28]。根据在不同相对压力下的氮气吸附量计算出来的比表面积、孔容积和平均孔径见表 4.1。在热解温度为 600 ℃下得到的 mPC-MgO-873 具有最好的结构特征(比表面积最大,孔容积最多)。和直接热解生物质本身获得的介孔碳相比(比表面积为 14.0 $m^2 \cdot g^{-1}$,孔容积为 0.067 $cm^3 \cdot g^{-1}$),mPC-MgO 的结构特征有了明显的提高,表明 $MgCl_2$ 的存在可以催化生物质热解碳化过程中孔隙的生成。

  mPC-MgO 的形貌通过扫描电子显微镜和透射电子显微镜来观测,结果如图 4.3 所示。SEM 图片显示 mPC-MgO 拥有多孔的结构,包含大量粗糙的片状碳,而 TEM 图片显示 MgO 纳米颗粒均匀地分布在碳片之上,其中 mPC-MgO-773 和 mPC-MgO-873 的颗粒尺寸主要分布在 35～69 nm 之间,而 mPC-MgO-973 的颗粒尺寸主要分布在 24～39 nm 之间,表明热解温度的升高有利于生成较小尺寸的 MgO 纳米颗粒。

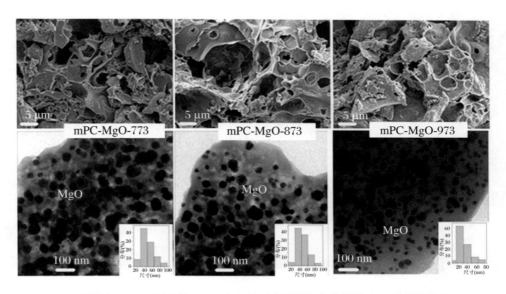

图 4.3 不同热解温度下制备的 mPC-MgO 的形貌观测和氧化镁纳米颗粒的粒径分布

4.1.4
介孔碳稳定的氧化镁纳米材料的 CO_2 捕集性能评价及机理

mPC-MgO 的 CO_2 捕集能力通过热重分析仪(DTG-60H/DSC-60,Shimadzu,Co.,Japan)进行测定。首先,将准确称量的一定量的(~10 mg)的 mPC-MgO 样品置于一个氧化铝盘中,然后在 100 mL·min^{-1} 的高纯氮气流中,在 300 ℃下脱气 30 min。接下来温度降低到设定值,气流转换为高纯 CO_2 气体(100 mL·min^{-1})来测定其 CO_2 捕集能力。为了研究温度的影响,实验过程中的温度从 50 ℃以 5 ℃·min^{-1} 的升温速率升高至 300 ℃,记录过程中 mPC-MgO 的质量变化数据。对于动力学分析,实验温度固定在 80 ℃,将脱气后的 mPC-MgO 置于 100 mL·min^{-1} 的 CO_2 气流中 90 min,记录过程中 mPC-MgO 的质量变化数据。对于 mPC-MgO 对 CO_2,N_2 和 O_2 气体的选择性分析,实验温度固定在 80 ℃,分别在高纯氮气和高纯氧气氛围(100 mL·min^{-1})下放置 60 min,记录过程中 mPC-MgO 的质量变化数据。对于 mPC-MgO 对 CO_2 的循环吸附实验,脱气后的样品先暴露于 CO_2 气流(100 mL·min^{-1})在 80 ℃反应 90 min,然后气流转换为氮气流(100 mL/min),温度升高至 300 ℃,反应 30 min,使 mPC-MgO 再生。这个过程重复多次,记录过程中 mPC-MgO 的质量变化数据,根据以下公式计算其 CO_2 捕集能力:

$$CO_2 \text{ 捕集能力(mol·kg}^{-1}) = \frac{(m_t - m_i)}{m_i M_w} \tag{4.5}$$

其中,m_i,m_t (kg)分别是初始和在 t 时的 mPC-MgO 质量,M_w (=0.044 kg·mol^{-1})是 CO_2 相对分子质量。

图 4.4(a)所示的是 mPC-MgO-873 在不同温度(50~300 ℃)下对 CO_2 的捕集能力。当温度从 50 ℃升高至 80 ℃时,mPC-MgO-873 的 CO_2 的捕集能力从 2.7 mmol·kg^{-1} 迅速升高至 5.2 mmol·kg^{-1},并达到最大值,当温度进一步升高至 300 ℃时,CO_2 的捕集能力迅速从 5.2 mmol·kg^{-1} 下降至 1.3 mmol·kg^{-1}。这种现象可以解释如下:mPC-MgO-873 对 CO_2 的捕集主要包括物理吸附和化学相互作用;对于物理吸附,其作用力随着温度的升高而迅速下降[29],而化学相互作用的作用力则是呈现出随着温度的升高先增加,然后再下降[30]。物理吸附和化学作用的协同作用导致了 mPC-MgO-873 在不同温度下对 CO_2 的捕集能力呈现出一个峰值。

图 4.4(b)所示的是温度为 80 ℃时 mPC-MgO 在不同时间下对 CO_2 的捕集能力变化。结果显示,mPC-MgO 对 CO_2 的捕集分为两个不同的阶段:第一阶段,一旦 mPC-MgO 暴露于 CO_2 氛围中时,呈现出一个迅速的质量增加过程,并且持续 20 min,然后增重过程变得缓慢直到停止。类似的第二阶段 CO_2 捕集动力学在其他相关文献中也有报道[30-31]。第一阶段 CO_2 捕集的动力学可以通过以下方程进行描述:

$$\ln\left(1 - \frac{q_t}{q_e}\right) = -kt \tag{4.6}$$

其中,$k(\min^{-1})$ 是速率常数,q_t 和 $q_e (mol \cdot kg^{-1})$ 分别是 CO_2 在时间 t 以及平衡时的捕集量。

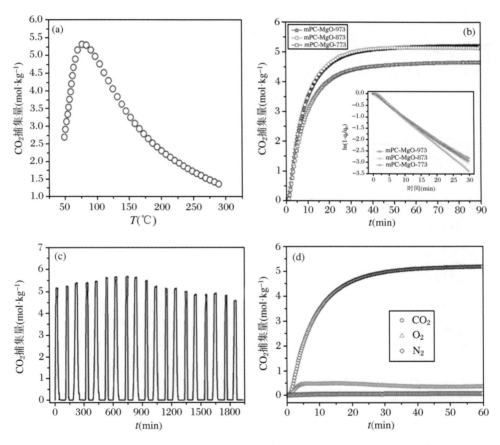

图 4.4 (a) 温度对 mPC-MgO-873 材料 CO_2 捕集能力的影响;(b) 不同温度制备的 mPC-MgO 对 CO_2 的捕集动力学;(c) mPC-MgO-873 材料对 CO_2 的循环吸附实验;(d) mPC-MgO-873 对 CO_2,N_2 和 O_2 的选择性评价

不同的 mPC-MgO 材料 CO_2 捕集动力学可以通过它们的速率常数进行比较,结果如图 4.4(b)插图所示。速率常数可以通过动力学方程线性拟合计算得

到,其值分别为 0.099 min^{-1},0.115 min^{-1} 和 0.102 min^{-1},相关系数(R^2)分别为 0.995,0.999 和 0.995,证实了一级动力学方程对 mPC-MgO 材料 CO_2 捕集动力学的描述是合理的,从而表明化学吸附是 CO_2 捕集过程的主要限速步骤。

对于 mPC-MgO 的实际应用来说,除了高的吸附性能和快的捕集速率之外,长期稳定性和可再生性也是一个重要的因素,同时,短的再生时间和低的再生稳定性也是必要的[32]。因此我们通过在循环吸附过程中 mPC-MgO-873 CO_2 吸附能力的变化来评价其可再生性和稳定性,结果如图 4.4(c)所示,当气体氛围由 CO_2 转换为 N_2 时,mPC-MgO-873 捕集的 CO_2 可以在不到 10 min 的时间内迅速脱附。吸附-脱实验重复了 19 次,发现在最初的 9 次循环中,材料对 CO_2 的捕集能力是不断增加的,这主要是因为捕集的 CO_2 可以起到对材料的活化作用。而随着循环次数的增加,材料对 CO_2 的吸附能力仅仅只有少量的降低,表明了 mPC-MgO 的长期稳定性。mPC-MgO 材料对 CO_2 的最大吸附量达到了 5.45 mol·kg^{-1},比其他很多的 MgO 基的 CO_2 捕集材料都要高很多,甚至可以和 CaO 基或者氨基的 CO_2 捕获材料相比(见表 4.2)。

表 4.2 不同 CO_2 捕获剂对 CO_2 捕集能力的比较

CO_2 捕集材料	最大捕集能力 (mol·kg^{-1})*	总循环次数	捕集温度 (K)	再生温度 (K)	参考文献
Al_2O_3 负载的 MgO	1.36	5	333	623	[36]
K_2CO_3 活化 MgO	1.98	17	748	748	[30]
介孔碳负载 MgO	2.09	6	298	623~723	[19]
介孔 MgO	2.27	3	373	1073	[37]
K_2CO_3 促进的 MgO	4.49	5	323	423~723	[31]
介孔碳稳定的 MgO 纳米颗粒	5.45	19	353	673	本研究
表面活性剂促进的胺类吸附剂	3.22	10	303	383	[7]
CaO-$Ca_{12}Al_{14}O_{33}$	~6.0	45	963	1163	[38]
介孔胺修饰的 SiO_2	6.97	—	298	—	[39]
Al_2O_3,CaO 和 MgO 复合物	7.73	65	923	1123	[40]
氨基功能化的介孔胶囊	7.90	50	348	383	[32]

注:* 以第一次循环捕集量所测值为准。

选择性是另外一个重要因素影响着 CO_2 捕获材料在实际应用中的性能,通常可以通过比较材料对 CO_2 和 N_2，O_2 这两种最主要的共存气体的捕集能力来表征，其计算公式如下[33-35]：

$$捕集能力 = \frac{CO_2 捕集能力}{N_2 或 O_2 捕集能力} \quad (4.7)$$

如图 4.4(d)所示，mPC-MgO 可以捕获的 N_2 和 O_2 分别为 0.045 mol·kg^{-1} 和 0.47 mol·kg^{-1}，远比其对 CO_2 的捕获能力低，mPC-MgO 对 N_2 和 O_2 的选择性分别为 112 和 11，这就显示，在 CO_2，N_2 和 O_2 共存的混合气体流中，mPC-MgO 可以高选择性的实现对 CO_2 的捕集。

mPC-MgO 对 CO_2 的高捕集能力可以归因于其多孔的结构和较大的比表面积。热解过程中形成多孔结构的机理可以解释如下：在热解的高温和高的升温速率条件下，水合氯化镁（$MgCl_2·6H_2O$）的脱水的分解过程可以同时促进生物质的分解产生挥发性物质，而挥发性物质的离去通常可以增加热解生物炭的孔隙和比表面积。同时，和氯化锌（$ZnCl_2$）类似，$MgCl_2$ 通常也对碳水化合物在高温下有很强的脱水能力，这可以改变生物质的分解途径，阻止热解过程中可能阻塞孔道的重质焦油的生成[41]，从而增加其表面积和孔隙率[42]。热解过程中形成的 MgO 等组分也可以同时作为模版促进孔隙的生成，类似的现象在热解其他含镁化合物制备多孔碳的文献中也有报道[43-44]。

除了比表面积大和多孔性之外，生物质热解过程中形成的功能基团在 CO_2 捕集过程中也起到明显作用。根据红外光谱分析（图 4.5(a)和表 4.3），在波数低于 1600 cm^{-1} 的区域，CO_2 捕集前后的材料都有 3 个主要的 FTIR 吸收峰，分别可以指认为 C—O 的伸缩振动，CH_2 的弯曲振动和 O—H 的弯曲振动。其中 O—H 的弯曲振动在 CO_2 捕集后从 1592 cm^{-1} 红移到了 1550 cm^{-1}，表明 OH 基团是 CO_2 过程中的主要参与基团之一。确实，由于氢原子和氧原子之间的电负性的差别巨大，OH 基团中的氢原子可以和 CO_2 中的氧原子形成氢键作用（O—H···O=C=O）（见图 4.5(d)）。已有文献证实了 CO_2 和 OH 基团之间的氢键作用，尽管比较弱，也能对 CO_2 的捕集能力有较明显的贡献[45]。此外，氢键的作用力是随着温度的升高不断下降的，因此也可以部分解释在高的温度下为什么 mPC-MgO 的捕集能力会有明显的下降。在 FITR 吸收光谱中波数高于 1600 cm^{-1} 的区域出现了两个新的红外吸收峰，其中位于 1631 cm^{-1} 处的吸收峰可以指认为是 C=O（羰基或者碳酸根）的伸缩振动，而在 2317 cm^{-1} 处的峰是 CO_2 的分子振动峰[39]，以上两个新的红外峰的出现可以证明物理吸附和化学作用都在 CO_2 的吸附过程中起重要的作用。

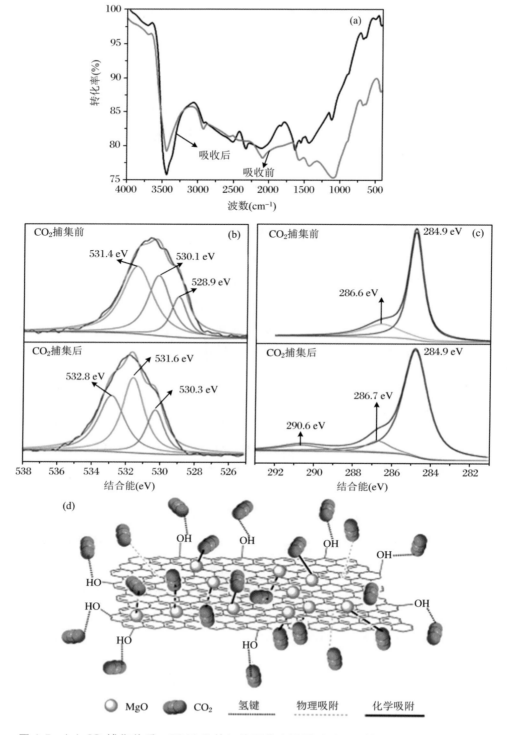

图 4.5 (a) CO_2 捕集前后 mPC-MgO 的红外吸收光谱图;(b) CO_2 捕集前后 mPC-MgO 的 XPS O 1s 图谱;(c) CO_2 捕集前后 mPC-MgO 的 XPS C 1s 图谱;(d) mPC-MgO 捕集 CO_2 的机理示意图

表 4.3　CO_2 捕集前后 mPC-MgO 的红外吸收光谱峰的位置归属分析

功能基团	位置（cm^{-1}）	
	CO_2 捕集后	CO_2 捕集前
O—H 或 O—Mg 伸缩振动	3439	3434
—CH 对称振动	2921	2916
CO_2 伸缩振动	—	2317
C=O 不对称振动	—	1631
O—H 弯曲振动	1592	1550
O—Mg 弯曲振动	1447	1442
C—O 伸缩振动	1119	1127
O—Mg 指纹弯曲振动	687	685
	633	637
	496	498

材料的表面化学状态也是影响其对 CO_2 捕集能力的主要因素，尤其是对于碳基的材料而言[46]。图 4.5(b) 和图 4.5(c) 所示的是 mPC-MgO 材料在 CO_2 捕集前后的 O 1s 和 C 1s 的 XPS 图谱，从中可知 mPC-MgO 材料表面化学状态的变化情况。其中 O 1s 的 XPS 图谱在 CO_2 捕集之前包含 3 个不同结合能的峰，分别为 O—C 和 O—H 基团（531.4 eV）[47]，在 MgO 晶面上 O^{2-} 离子（530.1 eV），以及在 MgO 晶体边缘或者角落的 O^{2-} 离子（528.9 eV）[48]。位于晶体边缘或者角落的 O^{2-} 离子通常比在晶面中的 O^{2-} 离子具有较强的碱性，可以更容易和 CO_2 在较低温度下反应[49-50]。因此，在 CO_2 捕集之后的 O 1s 图谱中，在 528.9 eV 处的峰消失了，反而在 532.8 eV 处出现了一个新的峰，证实了以上推测。这个机理也可以解释为什么 mPC-MgO 的 CO_2 捕集能力比大部分文献报道的 MgO 的 CO_2 捕集能力要高。图 4.5(c) 所示的是 C 1s 图谱，从中可知，在 CO_2 捕集前，C 1s 图谱只包含两个峰，分别是 C—C 和 C—O 基团，而在 CO_2 捕集之后，在结合能为 290.6 eV 处出现了一个新的峰，这是碳酸根（CO_3^{2-}）中的碳原子的 XPS 峰[51]，进一步证实了 CO_2 捕集过程中化学相互作用的存在（MgO + $CO_2 \longrightarrow MgCO_3$）。

在本节研究中，我们通过热解 $MgCl_2$ 负载的木屑生物质来制备介孔碳负载的 MgO 纳米颗粒材料，并将其用于 CO_2 的捕集，分析了其相关性能及其作用机理。根据实验结果，可以得出以下结论：

（1）热解过程，温度的升高有利于获得粒径较小的 MgO 纳米颗粒，但是，温度的升高也可能导致镁元素的挥发，降低其在热解生物炭中的含量。

(2) 不同热解温度下获得的介孔碳负载的 MgO 纳米颗粒材料对 CO_2 的捕集性能差别较大,其中热解温度在 600 ℃ 制备的材料具有最好的性能,其对 CO_2 的最大捕集量可以达到 5.45 mol·kg^{-1},高于大部分文献报道的 MgO 基的 CO_2 捕集材料。

(3) MgO 纳米颗粒材料对 CO_2 的捕集机理主要涉及物理吸附和化学相互作用,其中的物理吸附作用随着温度的升高而减弱,而化学相互作用主要包括氢键作用以及 CO_2 和 MgO 的化学反应,是对该材料捕集 CO_2 的主要贡献者。

4.2
生物质热解制备碳基磁性固体酸及催化性能评价

4.2.1
概述

酸性材料广泛应用于催化制备一系列的可再生能源(如生物柴油)和精细化学品(如糠醛或者糖类)[52-54]。早期应用最为广泛的酸性催化剂主要是液体矿物酸(HCl,H_2SO_4 和 H_3PO_4 等)[52,55]。但是这些液体矿物酸的使用通常存在许多不足之处,例如强腐蚀性、高成本、难以从均一的反应体系中分离,以及产生大量的废弃的副产物等,限制了其在很多领域的应用[56]。相对而言,固体酸材料,包括固体的 Brønsted 酸(如 SO_3H 功能化的碳基材料和聚合物)[57-60] 和 Lewis 酸(过渡金属氧化物:ZrO_2,WO_3,Nb_2O_5 等)[61-62],可以方便地从反应体系中分离,并且可以多次重复利用。因此固体酸材料可以取代液体矿物酸用于很多的酸催化反应中,包括酯化、水解和脱水反应,从而最大限度地降低这些反应对环境的副作用和成本[56,63-67]。

在所有的固体酸材料中,多孔碳基固体酸的合成和应用最为广泛[68-76]。

Nakajima 等[57]报道多孔碳基固体酸可以通过直接磺化部分碳化的有机物（糖类、纤维素、合成高分子等）来制备，用于催化水解 1,4-糖苷键或者酯交换合成生物油时，表现出良好的催化活性。Arancon 等[77]以玉米芯残渣为原料合成碳基固体酸材料用于催化酯化和酯交换反应，发现其具有十分优异的催化性能。和常规的固体酸材料相比，磁性固体酸材料可以在外部磁铁的存在下很方便地实现分离，尤其适用于黏稠和固体的反应体系[78-79]。最常见的合成磁性固体酸的方法是先制备一个含有磁性内核（如 Fe_3O_4）的多孔材料，然后进行磺化得到的。在之前的一个研究中，通过磺化聚苯乙烯包裹的 Fe_3O_4 可以得到具有高的酸强度（2.2～2.5 $mmol \cdot g^{-1}$）的磁性固体酸材料[80]。Zillillah 等通过磺化一系列包含磁性内核的多孔颗粒材料（如聚苯乙烯、聚甲基丙烯酸甘油酯和二氧化硅）制备了各种磁性纳米固体酸材料[81]。所有这些磁性固体酸材料都呈现出高的催化活性和良好的分离性能，但是仍然存在不足之处，例如磺化的磁性高分子通常热稳定性低且成本高，此外，制备这些磁性固体酸材料通常需要昂贵的化学试剂和复杂的合成步骤，这些缺点限制了其大规模的应用[82]。

在本节研究中，我们提出了一种快速、低成本、容易大规模生产的方法，以废弃生物质为前体，制备磁性多孔碳基固体酸材料。该方法主要涉及两个阶段，首先快速热解 Fe(Ⅲ)负载的生物质制备磁性介孔碳材料，然后再对其进行磺化，制备磁性多孔碳基固体酸材料。这个过程同时可以生产大量的生物油，可作为燃料和化学品原料。

4.2.2
磁性固体酸材料的合成过程及结构表征

固体酸材料的前体是 Fe(Ⅲ)负载的生物质，是通过使用生物质作为吸附剂从水中吸附 Fe(Ⅲ)得到的。吸附过程如下：将 10.0 g 的生物质加入 1000 mL 10 $mmol \cdot L^{-1}$ 的 $FeCl_3$ 溶液中，在常温常压下以 200 $r \cdot min^{-1}$ 的搅拌速度吸附 300 min。然后将混合物中的水分挥发出去，将固体残渣在 105 ℃干燥去除剩余的水分，过筛收集粒径小于 120 目的颗粒作为热解进料。磁性碳基固体酸的合成是通过快速热解 Fe(Ⅲ)负载的生物质获得磁性多孔碳前体，然后再经过磺化得到。快速热解的基本过程如 2.2.4 小节的描述基本一致。热解结束以后，生成的热解生物炭在 600 ℃下再经过 1 h 的碳化，然后冷却至室温，就得到了磁性多孔碳前体材料。将磁性多孔碳前体材料和浓硫酸混合（10 mL 浓硫酸配 1 g 固体)并超声 15 min,然后在 150 ℃ 条件下快速搅拌反应 10 h,使材料表面引入 SO_3H 基团。随后,反应混合物冷却至室温,然后缓慢加入一个含有 500 mL 去离子水的烧杯中。接下来将反应混合物过滤,并用去离子水清洗直到无硫酸根离子被检测到,然后用乙醇清洗几次去除水分,并在 80 ℃ 下干燥过夜,就此合成了磁性多孔碳基固体酸材料($MPC-SO_3H$)。作为对比,我们通过热解没有铁负载的生物质并磺化制备了一种没有磁性的碳基固体酸材料($PC-SO_3H$)。

固体酸中的碳和硫元素的含量通过一个高频红外碳硫分析仪(CS-600,LECO,USA)进行测定,而铁的含量是通过 X 射线荧光光谱仪来分析的(XRF-1800,Shimadzu Co.,Japan)。酸性位点是通过使用 NaOH 标准溶液(10 mmol·L^{-1})进行测定的,其基本方法如文献描述[83]。红外吸收光谱用于分析固体酸表面的功能基团。而固体酸的热稳定性通过热重分析仪进行表征(DTG-60H/DSC-60,Shimadzu Co.,Japan)。固体酸的结构特征通过氮气吸/脱附动力学来表征,测试之前,样品先在 120 ℃ 真空中脱气 10 h,然后在 −196 ℃ 下使用相应的测定仪器(Micromeritics Gemini apparatus,ASAP 2020 M + C,Micromeritics Co.,USA)进行表征。比表面积根据 BET 方法计算得到,总的孔容积通过在相对压力为 0.99 时吸附的氮气量来计算。固体酸的表面形貌通过透射电子显微镜(TEM,JEOL-2100F,Japan)和扫描电子显微镜(SEM,Sirion 200,FEI electron optics company,USA)来观测,并且通过一个耦合的 X 射线能谱仪(EDX,INCA energy,UK)来分析材料的微区组成。X 射线光电子能谱仪(ESCALAB250,USA)用于分析固体酸材料的表面化学状态。X 射线衍射仪(MXPAHF,Japanese Make Co.,Japan)用于分析固体酸材料的晶体结构特征及其变化。

图 4.6 所示的是 $MPC-SO_3H$ 材料的合成过程示意图。合成磁性多孔碳前体材料是固体酸材料制备的主要过程,主要包括以下几个步骤:首先在生物质的干燥过程中负载的 $FeCl_3$ 水解形成铁的水合氧化物(FeOOH),然后经过快速热解过程形成磁性多孔碳材料。在这个过程中,生物质中的组分发生分解产生的挥发性物质冷凝后可以形成生物油,而 Fe_2O_3 则被热解产生的还原性物质(如 H_2、CO、无定形碳等)还原生成 Fe_3O_4,具体反应过程如以下方程所示:

$$FeCl_3 + 3H_2O \longrightarrow Fe(OH)_3 + 3HCl \uparrow \qquad (4.8)$$

$$Fe(OH)_3 \longrightarrow FeO(OH) + H_2O \qquad (4.9)$$

$$6FeO(OH) + 4H_2 \longrightarrow 2Fe_3O_4 + 4H_2O \uparrow \qquad (4.10)$$

$$6FeO(OH) + 4C \longrightarrow 2Fe_3O_4 + 2CO \uparrow \qquad (4.11)$$

$$6FeO(OH) + 4CO \longrightarrow 2Fe_3O_4 + 4CO_2 \uparrow \qquad (4.12)$$

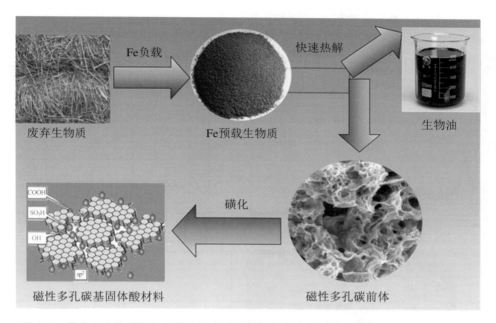

图 4.6 从废弃生物质制备磁性多孔碳基固体酸材料合成过程示意图

图 4.7(a)所示的是磁性多孔碳前体及其对照样品(无磁性碳)的 XRD 图谱:两个 XRD 图谱在衍射角度为约 23°处都有一个宽的峰,可以归因为那些芳香片状碳以随机的方式组成的无定形碳的衍射。而在谱图 A 中发现的在衍射角为 30.1°,35.7°,37.1°,43.2°,53.6°,62.7°和 74.9°的峰是 Fe_3O_4 的特征衍射(JCPDS 19-0629),这就证实了热解过程中 Fe(Ⅲ)最终转化为了 Fe_3O_4。

图 4.7(b)所示的是磁性多孔碳前体及无磁性碳的氮气吸/脱附曲线。对于磁性多孔碳前体,其吸/脱附曲线呈现出第三类型,在高的相对压力的区域有一个 H2 型的滞回环,表明材料中存在着大量不规则尺寸和形状的孔结构[84]。而对于无磁性碳对照样品,其吸/脱附曲线呈现出第二类型,并且没有明显的滞回环,表明该样品几乎没有孔结构。根据在不同相对压力下氮气的吸附量,通过 BET 方法计算出两个材料的比表面积,其中磁性多孔碳前体的比表面积为 391.7 $m^2 \cdot g^{-1}$,远远大于无磁性碳对照样品的比表面积(14.0 $m^2 \cdot g^{-1}$)。以上结果表明,铁的存在是可以催化生物质热解过程中形成多孔碳的结构的,这主要是因为一方面在热解的快速升温条件下,生物质负载的 $FeCl_3$ 发生脱水和分解,可以促进挥发组分如 CO_2、HCl、H_2O 和 CO 的产生(方程(4.8)~方程(4.12)),挥发组分的离去可以促进热解生物炭中孔隙的生长,从而增加其比表面积和孔容积,另一方面,$FeCl_3$ 在高温的情况下,可以改变碳水化合物聚合物的脱水分解路径,抑制热解过程中可能阻塞孔隙的重质焦油的产生,从而促进碳骨架中开放孔隙的形成,此外热解过程中形成的铁化合物(FeOOH,Fe_3O_4 等)也可以作为原

位催化剂和模版促进热解生物炭孔隙的生成[35]。类似的现象在文献报道的热解 FeCl₃ 浸渍的木屑或者碳凝胶生产多孔碳的实验中也有发现[85-86]。

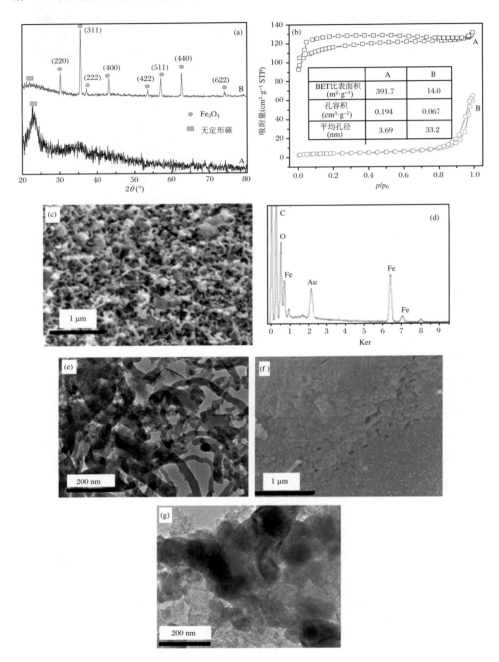

图 4.7 (a) 磁性多孔碳前体材料和无磁性碳对照样品的 XRD 图谱;(b) 磁性多孔碳前体材料和无磁性碳对照样品的氮气吸/脱附等温线;(c) 和 (d) 磁性多孔碳前体材料的 SEM 图片和其 EDX 微区结构分析;(e) 磁性多孔碳前体材料的 TEM 图片;(f)和(g) 无磁性碳对照样品的 SEM 和 TEM 图片

磁性多孔碳前体及无磁性碳的形貌通过 SEM 和 TEM 进行观察，SEM 的结果显示（图 4.7(c)）：磁性多孔碳前体的表面非常粗糙，包含有大量的线状结构。该结构在 TEM 图片（图 4.7(d)）中也有发现，并且在 TEM 图片中，还有不规则的多孔结构也被观察到了。而对于无磁性碳的对照样品，在其 TEM 和 SEM 图片中，线状结构和多孔结构都没有观察到，只是观察到了一个较为光滑的表面结构，这就证实了之前氮气吸/脱附曲线的结果。

然后将磁性多孔碳前体及无磁性碳对照样品都进行磺化反应制备固体酸材料。图 4.8(a)所示的是磁性固体酸材料的 XRD 图谱，从中可知，经过磺化过程之后，Fe_3O_4 的晶型并没有被明显的破坏。硫元素和铁元素在磁性固体酸中的分布情况通过 EDX 元素面扫描进行分析，我们可以发现，硫元素均匀地分布在整个磁性固体酸的表面（图 4.8(e)），表明磁性多孔碳前体材料得到了有效的磺化。如图 4.8(f)所示，铁元素也是较为均一的分布于磁性固体酸材料的整个表面上，也可以佐证 XRD 的结果，说明铁元素在磺化后并没有明显的流失，从而使得材料的磁性得以较好的保留。图 4.8(b)所示的是磁性固体酸材料的氮气吸/脱附曲线，根据计算得出的磁性固体酸的比表面积和孔容积分别只有 296.4 $m^2 \cdot g^{-1}$ 和 0.14 $cm^3 \cdot g^{-1}$，比磁性多孔碳前体材料稍微低一些，表明在磺化过程中，磁性多孔碳前体材料的孔结构被部分地破坏了。而 SEM 和 TEM 形貌表征则表明，磺化过程后，磁性固体酸几乎全部保留了其前体材料的基本形貌特征（线状多孔结构）。

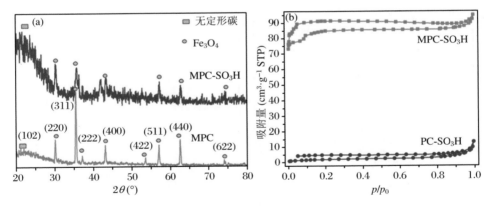

图 4.8 (a) 磁性多孔碳前体材料和磁性固体酸材料的 XRD 图谱；(b) 磁性固体酸材料和非磁性固体酸对照样品的氮气-吸/脱附曲线图；(c) 和(d) 磁性固体酸材料的 SEM 和 TEM 图片；(e) 和(f) 磁性固体酸材料的硫元素和铁元素的面分布图；(g) 和(h) 磁性固体酸材料的 C 1s 和 S 2p XPS 图谱

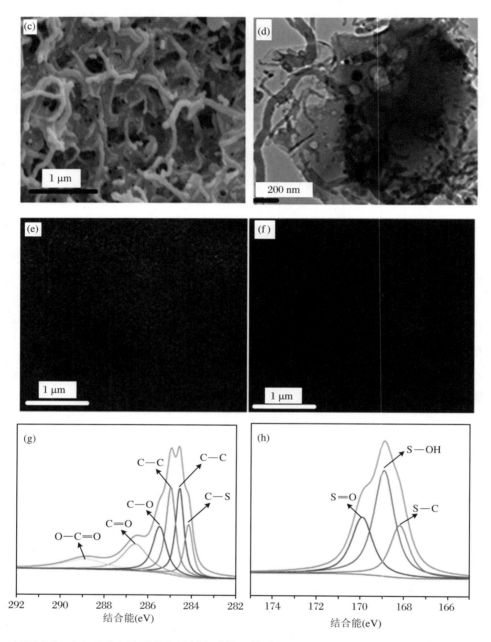

续图 4.8　(a) 磁性多孔碳前体材料和磁性固体酸材料的 XRD 图谱；(b) 磁性固体酸材料和非磁性固体酸对照样品的氮气-吸/脱附曲线图；(c) 和 (d) 磁性固体酸材料的 SEM 和 TEM 图片；(e) 和 (f) 磁性固体酸材料的硫元素和铁元素的面分布图；(g) 和 (h) 磁性固体酸材料的 C 1s 和 S 2p XPS 图谱

磁性固体酸材料的表面组成和化学状态通过 XPS 进行表征，其 C 1s 的 XPS 图谱包含 6 个不同结合能的峰(图 4.8(g))，这些峰分别可以归因为不同化学态的碳原子：C—S(284.1 eV)，C—C(284.6 eV 和 284.9 eV)，C—O(醇、酚或者醚类,285.4 eV),C=O(羰基或者喹啉类,286.7 eV)和 O=C—O(羧基,289.0 eV)。[87-88]而磁性多孔碳前体的 C 1s 图谱只包含两个峰,分别为 C—C 和 C—O,表明磺化过程不仅可以对碳材料引入磺酸基(形成 C—S 键),也可以对碳材料的表面修饰上更多的含氧功能基团(C=O 和 O—C=O 等),磁性固体酸的 S 2p 图谱包含 3 个不同的峰,分别为 S—C(168.1 eV),S—O(168.9 eV)和 S=O 键(169.9 eV)[57,89]。S 2p 的 XPS 图谱表明固体酸中的硫元素主要是来自和碳骨架相连的磺酸根[90]。因此我们可以通过硫元素的含量来估计固体酸中磺酸根的含量。如表 4.4 所示,磁性固体酸中的总的硫元素含量为 5.6 wt%,根据计算,可以得出磁性固体酸中的磺酸根的含量为 1.75 mmol·g^{-1},比其对照的非磁性固体酸的磺酸基含量(0.97 mmol·g^{-1})有明显的提高。而磁性固体酸和对照的非磁性固体酸的总酸性位点(根据 NaOH 溶液滴定法测定的)分别为 2.57 mmol·g^{-1} 和 1.26 mmol·g^{-1},表明磁性固体酸中除了磺酸根以外还有其他酸性位点的存在,这也被 XPS 和 FTIR 的结果所证实[91-92]。

表 4.4　固体酸材料的元素组成、磺酸基含量及总的酸性位点

	MPC-SO$_3$H	PC-SO$_3$H
元素组成(wt%)		
C	60.9	70.4
S	5.6	3.1
Fe	2.4	0.1
O*	31.1	26.4
SO$_3$H 含量(mmol·g^{-1})	1.75	0.97
总的酸性位点(mmol·g^{-1})	2.57	1.26

注:*通过差减法计算而得:$O\% = 100\% - C\% - S\% - Fe\%$。

4.2.3
磁性固体酸材料的催化性能评价

固体酸材料的催化活性可以通过 3 个典型的酸催化反应来进行表征,包括苯甲醇和乙酸的酯化反应(反应 A)、木糖的脱水反应(反应 B)和蔗糖的水解反应(反应 C)。

酯化反应是在一个无溶剂体系中进行的,其基本过程如下:将 0.100 g 的固

体酸催化剂加入一个 25 mL 的反应管中,再加入 0.108 g(1.0 mmol)的苯甲醇和 0.290 g(5.0 mmol)的乙酸,然后将反应物在磁力搅拌器中剧烈搅拌,在 90 ℃下反应。整个反应过程是在氮气氛围下进行的,并通过薄层色谱来监测反应过程。当反应结束时(薄层色谱没有苯甲醇检出),反应混合物先用 10 mL 的环己烷清洗,并使用强力磁铁将催化剂分离,剩下的液体部分先用 0.01 mol·L^{-1} 的碳酸氢钠溶液去除未反应的乙酸,然后用无水硫酸镁干燥,最终产物(乙酸苯甲酯)通过在减压蒸馏的情况下去除环己烷溶剂而得,产物的结构是通过和标准品乙酸苯甲酯在气相色谱的出峰位置进行比对而确定的。分离的催化剂用乙酸乙酯清洗几次,然后在 60 ℃干燥过夜回收再利用。

对于木糖的脱水反应,是将 0.150 g 左旋木糖(1 mmol),0.500 g 的 DMSO 和 0.100 g 的酸催化剂混合于 1 个 25 mL 的反应管中。然后将反应混合物在恒温油浴锅中,于 150 ℃下剧烈搅拌反应 300 min。反应结束过后,将混合物冷却至室温并分析产物的生成和反应物的消耗情况。生成的糠醛的浓度通过气相色谱进行测定,而未反应完的木糖则通过间苯三酚比色法,使用紫外-可见分光光度计来分析,基本方法见相关文献报道[93]。木糖的转化率和糠醛的产率通过以下公式进行计算:

$$木质素转化率(\%) = \frac{初始木糖 - 未反应木糖}{初始木糖} \times 100\% \quad (4.13)$$

$$糠醛产率 = \frac{糠醛含量}{初始木糖} \times 100\% \quad (4.14)$$

对于蔗糖的水解反应,将 0.342 g(1 mmol)的蔗糖、20 mL 水和 0.100 g 的酸催化剂加入一个 100 mL 的圆底烧瓶中,然后在 80 ℃下恒温水浴反应 150 min。反应结束后,混合物冷却至室温并分析产物的生成。水解产物葡萄糖和果糖采用 3,5-二硝基水杨酸比色法测定[94]。未反应完的蔗糖通过在 6 mol·L^{-1} 的盐酸溶液进一步水解后采用同样的方法测定,蔗糖的转化率和水解产物的产率通过以下公式进行计算:

$$蔗糖转化率(\%) = \frac{初始蔗糖 - 未反应蔗糖}{初始蔗糖} \times 100\% \quad (4.15)$$

$$水解产物产率 = \frac{葡萄糖和果糖含量}{初始蔗糖} \times 100\% \quad (4.16)$$

磁性固体酸材料的催化性能通过 3 个典型的酸催化反应来表征(图 4.9),分别为:苯甲醇和乙酸的酯化反应(A);木糖脱水生成糠醛的反应(B)和蔗糖水解生成葡萄糖和果糖的反应(C),其基本反应方程式如下所示:

$$\text{苯甲醇} + \text{乙酸} \xrightarrow[363\,K]{\text{酸催化剂}} \text{乙酸苯甲酯} \qquad A$$

$$\text{木糖} \xrightarrow[423\,K\ DMSO]{\text{酸催化剂}} \text{糠醛} \qquad B$$

$$\text{蔗糖} \xrightarrow[353\,K\ H_2O]{\text{酸催化剂}} \text{葡萄糖} + \text{果糖} \qquad C$$

对于苯甲醇和乙酸的酯化反应，磁性固体酸表现出很高的催化活性，可以使得反应产物乙酸苯甲酯的产率达到 93%，远比其对照样品无磁性固体酸的催化产率(63%)高，甚至高于常用的液体矿物酸-浓硫酸的催化产率(89%)。而对于反应 B，这是一个从生物质生产糠醛的最基本的反应，在无催化剂存在的情况下，只有 35% 的木糖被转化，而生成糠醛的产率几乎为 0，而在磁性固体酸的存

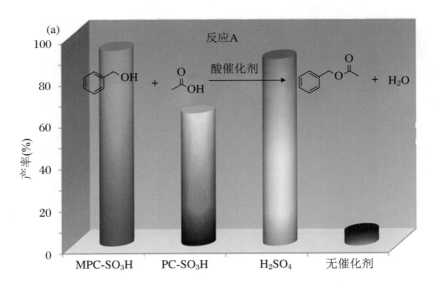

图 4.9　(a) 磁性固体酸材料和其他酸性催化剂对于苯甲醇和乙酸酯化反应的催化性能比较；(b) 磁性固体酸材料和其他酸性催化剂对于木糖脱水反应的催化性能比较；(c) 磁性固体酸材料和其他酸性催化剂对于蔗糖水解反应催化性能比较；(d) 磁性固体酸材料循环使用性能

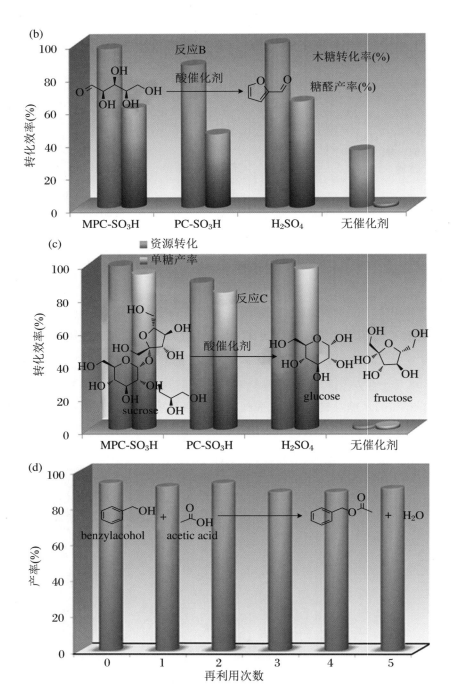

续图 4.9 (a) 磁性固体酸材料和其他酸性催化剂对于苯甲醇和乙酸酯化反应的催化性能比较;(b) 磁性固体酸材料和其他酸性催化剂对于木糖脱水反应的催化性能比较;(c) 磁性固体酸材料和其他酸性催化剂对于蔗糖水解反应催化性能比较;(d) 磁性固体酸材料循环使用性能

在下,糠醛的产率可以达到61%,远高于对照样品无磁性固体酸的催化产率(45%),但是稍微比浓硫酸的催化产率低一些,表明了该磁性固体酸催化剂的优越的催化性能[95-96]。对于蔗糖的水解反应,在没有催化剂存在的情况下,蔗糖的水解率几乎为0,而在磁性固体酸的作用下,蔗糖几乎可以100%的水解,生成葡萄糖和果糖的产率为94%,和浓硫酸的催化效率几乎相当,但是明显高于无磁性固体酸的催化效率。此外,磁性固体酸催化蔗糖水解的反应速率也很快,如图4.10所示,在1 h以内,蔗糖的水解率就可以达到90%以上。以上结果表明,合成的磁性固体酸材料对不同的酸催化反应都具有较高的催化效率。

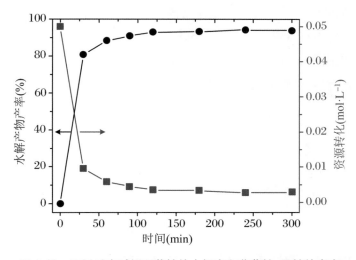

图4.10　不同反应时间下蔗糖的水解率和葡萄糖、果糖的产率

对于磁性固体酸而言,其相对液体矿物酸最主要的优点就是易分离回收利用性。因此,在本研究中,反应结束后磁性固体酸材料通过一个外部的强力磁铁分离后再将其重复应用于催化乙酸和苯甲醇的酯化反应,结果显示,经过5次重复利用后,苯甲醇乙酸酯的产率仍然能达到90%以上,并且其磁性也得以保留。XPS结果显示(表4.5),5次使用后的磁性固体酸表面元素组成也基本保持不变,表明了该磁性固体酸材料优异的催化稳定性能。

表4.5　磁性固体酸材料循环5次使用前后的表面元素组成和化学状态的XPS分析结果

	使用之前的磁性固体酸		5次使用以后的磁性固体酸	
	峰点结合能(eV)	原子百分比(%)	峰点结合能(eV)	原子百分比(%)
C 1s	284.8	70.83	284.8	74.06
O 1s	533.0	26.58	532.3	23.57
S 2p	169.1	1.32	168.9	1.23
Fe 2p	711.9	1.27	713.0	1.14

在本节研究中,我们通过快速热解 $FeCl_3$ 负载的生物质制备磁性多孔碳前体,再将其磺化合成多孔磁性固体酸材料。根据实验结果,可以得出以下结论:

(1) 热解过程中,生物质负载的 $FeCl_3$ 在热解的还原气氛下,经过脱水和还原过程形成 Fe_3O_4 晶体,并且在磺化过程中 Fe_3O_4 晶体没有破坏,从而赋予固体酸材料磁性。

(2) 在热解过程中,$FeCl_3$ 的存在可以催化促进形成多孔碳质材料,其比表面积为 391 $m^2 \cdot g^{-1}$,经过磺化过程以后,其比表面积稍有降低,但是仍然达到 296 $m^2 \cdot g^{-1}$。

(3) 合成的磁性固体酸具有很高的酸强度,其酸性基团包括磺酸基团和羧基等,总的酸性位点含量为 2.57 $mmol \cdot g^{-1}$。

(4) 磁性固体酸对苯甲醇和乙酸的酯化反应、木糖脱水生成糠醛的反应以及蔗糖水解生成葡萄糖和果糖的反应都具有很高的催化活性,与浓硫酸的催化性能相当。

4.3 快速热解碳化法从富氮湿地植物生物质制备氮掺杂多孔碳材料及其电化学储能性能研究

4.3.1 概述

化石燃料的过度消耗引发了全球性的环境问题,诸如能源危机和气候变化等,使得人们迫切需要寻找低成本、洁净可再生的替代能源。然而,大部分的洁净可再生能源,例如风能、太阳能和生物质能,都是间歇性的,无法方便使用和储存[97]。超级电容器是一类电子器件，提供了一个理想的途径用于快速储存间歇性的可再生能源[98-99]。由于其较高的功率和能量密度，超长的循环使用寿命以及优异的可逆性，超级电容器已经广泛应用于家用电子产品、工业设备和移动电子系统中[100-101]。

超级电容器的能量储存主要通过两个不同的途径：即电双层电容（EDLC）和赝电容。电双层电容器主要是基于电解液离子在电极材料表面的扩散和转移[102]，而赝电容器主要是基于电极材料的法拉第氧化还原反应[103]。尽管赝电容器，例如导电聚合物[104]、金属氧化物材料[105]通常具有较高的比电容，然而它们的稳定性和耐用性通常很差，这就限制了赝电容的商业化应用。相反，优异的循环耐用性使得电双层电容器几乎占据了100%的商用电容器的应用[106]。碳质材料，包括活性炭[107]、碳纳米管和纳米线[108-109]以及石墨烯等[110-111]，是目前应用最为广泛的电双层电容器材料[112-122]。它们通常具有较大的比表面积、多孔的结构以及高度稳定性，使其具有高的电容和优良的耐用性。然而碳材料的复杂和无规则的多孔结构限制了电子传递的效率并且导致材料的导电性能降低，从而极大地限制了电容器的功率和能量密度[99]。因此，需要通过其他的方法来提高碳基的电双层电容器的功率和能量密度。例如Bordjiba等通过在碳纳米管的骨架上沉积高比表面积的碳气凝胶，合成了一个具有很高功率和能量密度的超级电容器材料，其中的碳纳米管不仅可以作为一个结构骨架，也能促进电荷的传递过程[123]。

除了在导电性高的碳骨架结构中负载一些多孔碳，在碳骨架结构中掺杂一些杂原子引入赝电容特性是提高电双层电容器性能的另一个主要方法[124-126]，通常在碳骨架中掺入氮原子，这主要是由于氮原子不仅可以改善碳材料的亲水性和导电性，同时也能保持它们优异的循环稳定性[127-129]。常规的氮原子掺杂方法包括在高温条件对碳材料使用氨气处理[130]，以及直接碳化含氮的高分子聚合物，例如三聚氰胺、聚丙烯腈和聚乙烯吡啶等[131-133]。尽管这些方法可以合成不同比表面积和氮含量的碳材料，但是通常需要复杂的合成步骤和一些昂贵的试剂和设备，这限制了其大规模的商业化生产。

水生植物生物质是一种广泛存在的可再生资源，可作为一个理想的原材料用于合成高附加值的碳材料[134]。另一方面，水生植物具有很强的氮富集能力[135]，可以修复氮含量过高的富营养化水体。这些湿地植物生物质通常具有较高的氮含量[136]，可以直接用作原材料合成氮掺杂的多孔碳材料，容易实现低成本大规模的生产。在本节研究中，我们采用一种具有很强氮富集能力、生长迅速的水生植物——蒲草作为原料，通过快速热解碳化的方法来合成氮掺杂的多孔碳材料并将其用于超级电容器电极材料和测试其多种性能。

4.3.2 材料合成过程中氮元素的迁移转化过程及其机理

蒲草生物质采集自南淝河,生物质收割之后先清洗去除灰尘等杂质,然后在105 ℃下干燥过夜。干燥过的生物质粉碎过筛收集粒径小于 0.15 mm 的颗粒。将 5.0 g 干燥的水生植物生物质放置于一个自由落体固定床热解反应器中,首先进行一个预碳化过程,获得热解生物炭材料,热解过程和之前文献中的描述是一致的[136]。产生的固体生物炭通过 KOH 高温活化进一步增加其比表面积和孔隙结构。具体活化过程表述如下:将固体生物炭和粉末 KOH 均匀地混合(混合比为 1∶4),然后以 3 K·min^{-1} 的升温速率下在 100 mL·min^{-1} 的氩气氛围下加热至 700~900 ℃,并保持恒温 2 h,然后冷却至室温。将活化后的固体用 10 wt%的盐酸溶液清洗多次,去除其中的无机化合物,再用去离子水洗涤直到 pH 中性。最后将样品在 105 ℃的马弗炉中干燥就获得了最终的氮掺杂的多孔碳材料(简写作:N-pC-T,T 为活化温度)。

蒲草生物质首先经历一个在 500 ℃下的快速热解过程,获得氮富集的热解生物炭材料。在这个过程中,生物质中的小部分含氮组分(主要是植物氨基酸和生物碱等)在极高的升温速率(>500 K·min^{-1})和温度下快速分解,转化成一些小分子的含氮杂环化合物(如吡啶、吡咯及其衍生物等)并挥发至生物油中[52,137]。其他的氮则将嵌入碳骨架中而保留在热解生物炭中,这部分的氮占到生物质中总氮含量的 50%以上。热解生物炭中的氮含量为 3.99 wt%,远远高于其在原始生物质中的含量(2.93 wt%)。生物质热解前后的氮元素化学状态的变化通过 XPS 表征,结果如图 4.11 所示。热解之后,在原始生物质中发现的 N 1s 图谱中的两个峰:402.2 eV(酰胺 N—C═O)和 405.2 eV(硝酸盐)消失了,而一个新的峰出现在了 398.6 eV。这个峰可以归结为 pyridinic N(在碳骨架中的氮元素),证明了在热解过程中氮元素是可以嵌入碳骨架中的。

尽管在快速热解过程,氮元素的含量增加并且可以嵌入到碳骨架中,然而所得的热解生物炭的比表面积很低,孔隙也不发达,并不适合作为超级电容的电极材料,需要使用 KOH 活化增加其比表面积和孔隙结构。KOH 活化促进材料比表面积增加和孔隙形成的机理可以用反应方程式(4.17~4.20)表示。首先热解生物炭上面的功能基团在高的活化温度下进一步分解释放出挥发性物质(H_2O,

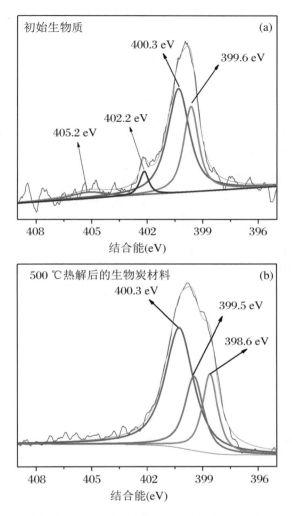

图 4.11　蒲草生物质热解前后 N 1s XPS 图谱

CO_2 和 CO 等），其中的 CO_2 可以被 KOH 捕获而形成 K_2CO_3。而在高温下，KOH 本身可以和碳发生反应，生成挥发性的 CO 和 H_2，并被还原成金属钾。在温度达到 774 ℃ 以上时，金属钾也将发生气化而挥发出去[138]。这些气体物质的挥发将会在材料上产生大量的孔隙结构。而随着活化温度的升高，之前生成的 K_2CO_3 也将进一步和碳发生反应，生成更多易挥发的其他物质（CO 和金属钾等），继续增加孔隙和比表面积[139]。

$$2KOH + CO_2 \longrightarrow K_2CO_3 + H_2O\uparrow \tag{4.17}$$

$$2C + 2KOH \longrightarrow 2CO\uparrow + 2K\uparrow + H_2\uparrow \tag{4.18}$$

$$K_2CO_3 + C \longrightarrow K_2O + 2CO\uparrow \tag{4.19}$$

$$K_2O + C \longrightarrow 2K\uparrow + CO\uparrow \tag{4.20}$$

表 4.6 所示的是氮掺杂多孔碳材料（N-pC）的元素组成和结构特征（比表面

积、孔容积和平均孔径等），从中可知，随着活化温度的升高，N-pC 材料的氮含量从 2.2 wt%下降至 1.6 wt%，表明在活化过程中，热解生物炭中的部分氮元素将会进一步挥发[140]。图 4.12 所示的是不同活化温度获得的 N-pC 材料的 XPS 扫描图谱，我们将 XPS 分析得到的表面元素含量比与元素分析仪测定的结果进行对比，发现了一些有趣的现象。尽管这两种方法测定的 N/C 和 O/C 都是随着活化温度的升高而降低的，但是 XPS 测定的 N/C 和 O/C 的值比元素分析仪测定的值要大很多，表明材料表面的 N 和 O 的含量明显高于其整体含量，这对材料作为超级电容器的电极材料是十分必要的，因为只有在材料表面的杂原子对其赝电容能力有贡献，并且可以增加其导电性[141-142]。同时 N-pC 的 N 1s 图谱（图 4.13）表明更多的吡啶氮（结合能：398.5 eV）和季铵氮（结合能：401.3 eV）在活化过程中形成了，而这种化学形态的 N 元素通常十分稳定，并且对电容性能有很大的帮助[133]。

表 4.6　不同活化温度下合成的 N-pC 的元素组成和结构特征

	N-pC-973	N-pC-1073	N-pC-1173
元素组成（wt%）①			
C	79.7	88.1	92.2
H	2.3	1.2	1.0
N	2.2	2.1	1.6
O	15.8	8.6	5.2
O/C（mol/mol）	0.15	0.073	0.042
N/C（mol/mol）	0.024	0.020	0.015
表面元素组成（atom%）②			
C	81.15	86.44	89.83
N	2.49	2.24	1.59
O	16.36	11.32	8.58
O/C（mol/mol）	0.20	0.13	0.096
N/C（mol/mol）	0.031	0.026	0.018
结构特征			
BET 比表面积（$m^2 \cdot g^{-1}$）	2361.1	3061.8	3110.7
总孔容积（$cm \cdot g^{-1}$）	1.67	2.09	1.79
平均孔径（nm）	2.15	2.73	3.03

注：① 通过元素分析仪测定。
　　② 通过 XPS 测定。

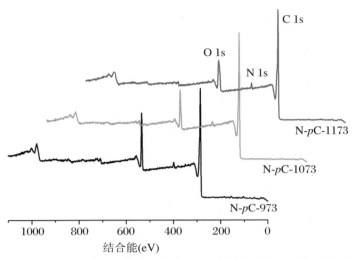

图 4.12　不同活化温度下合成的 N-pC 材料的 XPS 全谱扫描图谱

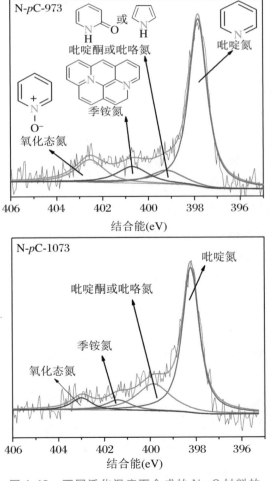

图 4.13　不同活化温度下合成的 N-pC 材料的 N 1s XPS 图谱

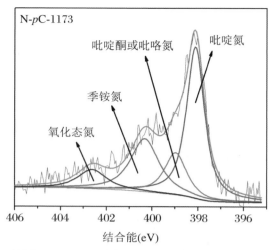

续图 4.13 不同活化温度下合成的 N-pC 材料的 N 1s XPS 图谱

4.3.3
材料的结构表征

氮掺杂多孔碳材料的比表面积和孔隙结构通过氮气吸附/脱附等温线在 -196 ℃下,采用粉末吸附仪(Micromeritics Gemini apparatus,ASAP 2020 M+C,Micromeritics Co.,USA)来表征。其比表面积通过 BET 方法计算,而它们的孔容积则通过在相对压力为 0.99 时吸附的氮气的量来计算的。材料的拉曼光谱通过一个激光拉曼光谱仪(LabRamHR,HORIBA Jobin Yvon. Co.,France)来分析,其激光源的波长为 514 nm,功率为 25 mW。材料的主要元素组成(C,H,N 和 O)通过元素分析仪(VARIO EL Ⅲ,Elementar Inc.,Germany)进行测定。材料的表面元素组成和化学状态通过 XPS 在恒能量分析模式(Constant Analyzer Energy,70 eV 全谱扫描,20 eV 高分辨谱)进行表征。表面形貌通过扫描电镜(SEM,Sirion 200,FEI electron optics company,USA)和透射电镜观察。

图 4.14(a)所示的是不同活化温度下制备的 N-pC 材料的氮气吸附/脱附等温线,对于 N-pC-973,根据 IUPAC 的分类,它的等温线呈现出 Ⅱ 和 Ⅳ 的过渡态[143],并在相对压力在 0.5~0.8 之间有一个小的滞回环。而 N-pC-1073 和 1173 的等温线则呈现出典型的 Ⅳ 型,并且在相对压力在 0.4~0.9 之间有一个清晰的滞回环,表明在这些材料中微孔(孔径<2 nm)和介孔(孔径=2~50 nm)

并存[27]。材料的 BET 比表面积、孔容积和平均孔径归纳于表 4.6，从中可知，N-pC 材料的比表面积从活化温度为 700 ℃时的 2361.1 m^2·g^{-1} 迅速升高至 800 ℃时的 3061.8 m^2·g^{-1}，然后随着温度的进一步升高至 900 ℃时，其比表面积只有稍微的增加到 3110.7 m^2·g^{-1}。如此高的比表面积可以提供足够的电极材料-电解液表面用于容纳电荷和电解液离子，对其电容性能起到至关重要的作用[144]。

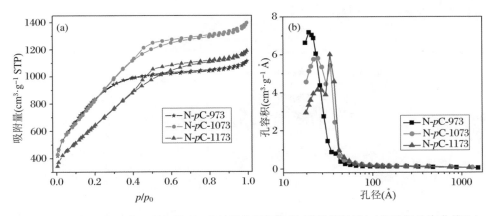

图 4.14　不同活化温度下制备的 N-pC 材料的氮气吸附-脱附等温线(a)和孔径分布曲线(b)

图 4.14(b)所示的是 N-pC 材料的孔径分布，对于 N-pC-973，其孔径分布曲线只有一个峰值出现在 2.1 nm 处，而对于 N-pC-1073 和 N-pC-1173 的孔径分布曲线都有两个峰值分别出现在 1.9 nm 和 3.4 nm 处，这就证实了在这些样品中微孔和介孔的共存。N-pC-973，N-pC-1073 和 N-pC-1173 材料的平均孔径分别为 2.15 nm，2.73 nm 和 3.03 nm，表明了高的活化温度有利于材料孔径的生长，证实了之前的机理推断。材料的孔隙微结构也可以直接通过 SEM 和 TEM 进行观察，从图 4.15 可知，所有的 N-pC 样品都含有很多不规则形状的孔隙结构，而且 N-pC-1173 的孔隙尺寸明显比 N-pC-1073 和 N-pC-973 的要大。

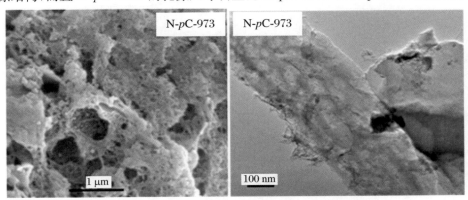

图 4.15　不同活化温度下制备的 N-pC 材料的 SEM(a)和 TEM(b)图片

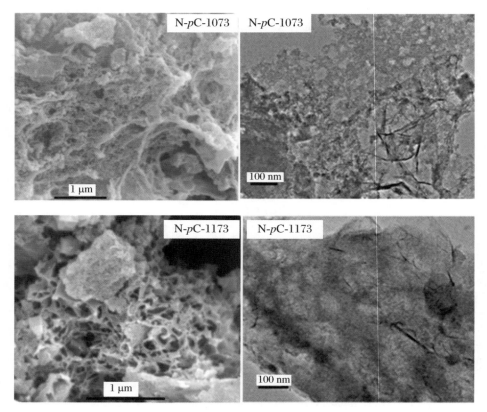

续图 4.15 不同活化温度下制备的 N-pC 材料的 SEM(a) 和 TEM(b) 图片

N-pC 材料的拉曼光谱图可以观察到碳元素的 G 峰和 D 峰(图 4.16),这两个峰分别反映的是碳材料的石墨化程度和混乱度[145-147]。N-pC-973,N-pC-1073 和 N-pC-1173 材料的 G 峰和 D 峰的比值(I_G/I_D)的比值分别为 1.03,

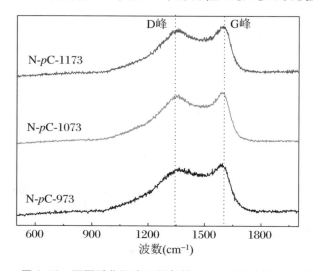

图 4.16 不同活化温度下制备的 N-pC 材料的拉曼光谱

1.04 和 1.09，表明活化温度的升高有利于提高碳材料的石墨化程度，而石墨化程度的提高对碳材料导电性的提升是十分必要的，从而也可以一定程度上改善材料的电容性能。

4.3.4
材料的电容性能评价及其作用机制探索

材料的电化学性能是在常温三电极体系下，采用 CHI 760D 电化学工作站进行测试的，其中 Ag/AgCl 电极用作参比电极，铂丝用作电极，0.5 mol·L^{-1} 的 K$_2$SO$_4$ 用作电解液。工作电极通过在泡沫镍片上负载含有 80% 的活性材料、10% 的导电介质（炭黑）和 10% 的黏合剂（聚四氟乙烯甲醇悬浮液）的糊状固体，然后在 80 ℃ 下干燥 30 min，然后将初步干燥的泡沫镍电极压紧，并在 100 ℃ 烘箱中干燥过夜。每个工作电极的活性材料负载约为 3.0 mg，面积为 1.0 cm^2。循环伏安（CV）曲线在电压范围为 0~0.8 V（vs Ag/AgCl）扫描得到，其扫描速率在 2~200 mV·s^{-1} 之间。根据 CV 曲线，通过以下公式可以计算出材料的比电容[148]：

$$C = \frac{1}{mv(V_f - V_i)} \int_{V_i}^{V_f} I(V) dV \tag{4.21}$$

其中，C（F·g^{-1}）是根据活性材料负载量计算出来的比电容，I（A）是电流密度，V（V）是电压，v（V·s^{-1}）是扫描速率，m（g）是每个电极上的活性材料的负载量，V_i 和 V_f 分别是扫描初始和最终的电压。恒电流充放电实验是在 0.5~6.0 A·g^{-1} 的充电电流下进行测试的，材料的放电电容根据以下公式计算[149]：

$$C_m = \frac{I \Delta t}{m \Delta V} \tag{4.22}$$

其中，I（A）是放电电流，Δt（s）是放电时间，m（g）是工作电极上负载的活性材料，ΔV（V）是充放电的电压范围，C_m（F·g^{-1}）是放电电容。电化学阻抗谱在开路电压下，频率在 10~100 kHz 之间进行测定。

图 4.17 所示的是 N-pC 电极材料在不同扫描速率下的循环伏安曲线（CV）。不同活化温度下制备的 N-pC 电极材料的 CV 曲线都呈现出粗略的矩形，表明材料电容的主要贡献来自 EDLC[140]。在扫描速率较低时（<50 mV·s^{-1}），在 CV 曲线电压为 0.2~0.4 V 处出现了一个小的突起峰，这可以归结为电极材料表面含氮功能基团发生氧化还原而产生的赝电容特性[150-151]。随着扫描速率达到 100 mV·s^{-1} 以上时，这个小的突起峰将会消失，

表明含氮基团的氧化还原具有相对缓慢的反应动力学[152]。此外，随着扫描速率增加到 100 mV·s^{-1} 以上，CV 曲线的矩形将会变得不规则（图 4.17），这主要是因为在高的扫描速率下，电解液离子和电极材料之间的接触将变得不充分[129]。图 4.17(d)归纳了根据不同扫描速率下 CV 曲线计算出来的 N-pC 样品的比电容：N-pC-1073 的比电容值最大可以达到 257 F·g^{-1}，明显高于其他两个 N-pC 样品（N-pC-973 为 232 F·g^{-1} 和 N-pC-1173 为 140 F·g^{-1}）。257 F·g^{-1} 的电容值也使得 N-pC-1073 比大部分文献报道的氮掺杂碳材料都要好（表 4.7）。根据方程 $C = \varepsilon_r \varepsilon_0 A/d$（其中 ε_r, ε_0 分别为电解液的介电常数和真空介电常数，A 是电极材料的比表面积，d 是电荷的分离距离）[153]，电极材料的比表面积越大，其电容越大。相比 N-pC-973 和 N-pC-1073，尽管它们的氮含量十分接近，但是 N-pC-1073 的比表面积远高于前者，导致了 N-pC-1073 具有较高的比电容。而对比 N-pC-1073 和 N-pC-1173，尽管它们的比表面积相差不大，可 N-pC-1073 具有较高的氮含量（2.1 wt%对 1.6 wt%），而碳材料骨架中掺杂的氮原子可以改变碳原子的电子结构，同时还能通过氧化还原反应产生赝电容特性[103]，从而使得 N-pC-1073 具有较高的比电容。此外，图 4.17(d)也显示，随着扫描速率的

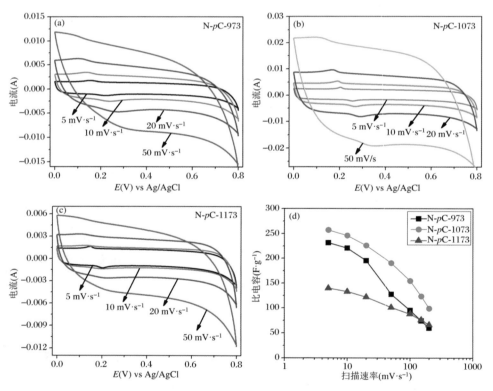

图 4.17 (a)～(c)不同温度下制备的 N-pC 在低扫描速率（＜50 mV·s^{-1}）下的循环伏安曲线图；(d) 不同扫描速率下 N-pC 材料的比电容

增加，N-pC 材料的电容将会持续的下降，这种下降是广泛存在于 EDLC 电极材料中，并且对于多孔碳基超级电容材料来说是不可避免的，其主要原因是在高的扫描速率下，没有充分的时间用于电解液离子的扩散和转移[152,154]。确实，如图 4.18 所示，在高的扫描速率下，CV 曲线的矩形将会明显的变形，这个是 EDLC 电极材料表面电荷的扩散和转移被限制的主要特征[154]。

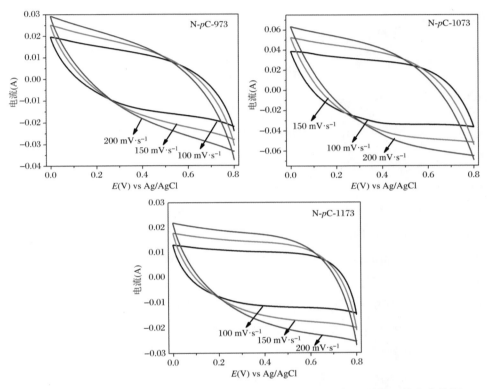

图 4.18　不同温度下制备的 N-pC 在高扫描速率（>50 mV·s^{-1}）下的循环伏安曲线图

表 4.7　N-pC 材料和其他文献报道的氮掺杂的多孔碳材料的电容比较

前体	电极材料	氮含量	比表面积 (m^2·g^{-1})	电容 (F·g^{-1})	参考文献
蚕丝肽	KOH 活化氮杂碳	0.89 atom%	3075	52	[155]
碳纳米管和聚丙烯腈	氮掺杂碳与碳纳米管复合物	9.2 wt%	157	100	[156]
三聚氰胺、甲醛与碳纳米管聚合物	氮掺杂纳米管	12.0 wt%	393	126	[157]
聚丙烯腈与聚丁基丙烯酸脂	富氮纳米碳	0.9 wt%	2570	173	[133]

续表

前体	电极材料	氮含量	比表面积 $(m^2 \cdot g^{-1})$	电容 $(F \cdot g^{-1})$	参考文献
碳纳米线与吡咯	氮杂碳纳米线	4.0%（原子百分数）	563	202	[129]
聚乙烯吡啶与聚丙烯腈	介孔氮杂碳材料	7.0 wt%	2000	202	[158]
三聚氰胺与甲醛树脂	介孔氮杂碳材料	10.3 wt%	1330	211	[159]
聚吡咯	氮杂微孔碳球	2.2 wt%	1080	240	[160]
废弃水生植物生物质	氮杂多孔碳材料	2.1 wt%	3061	257	本工作
二氰二胺	介孔氮杂碳材料	13.1 wt%	586	262	[161]
壳聚糖与葡萄糖胺	氮杂水热碳	4.4 wt%	571	300	[103]

为了进一步评价材料的电容性能，我们在三电极体系下，电压范围在 0~0.8 V 之间对材料进行恒电流充放电实验。如图 4.19（a~c）所示，在低的充电电流密度下，材料的充放电曲线呈现出明显的弯曲，这是赝电容的主要特征[162]，而随着充电电流密度的增加，充放电曲线的弯曲也变得不明显，直到消失。这种现象可以解释为表面含氮功能基团的氧化还原反应产生赝电容特性，而这些氧化还原反应在高充电电流密度下将受到严格的限制，从而降低了赝电容对材料电容性能的贡献。

能量和功率密度是评价材料电容性能的重要参数，可以通过以下公式进行计算[163]：

$$E = \frac{1}{2} \times C_m \times (\Delta V)^2 \tag{4.23}$$

$$P = \frac{E}{\Delta t} \tag{4.24}$$

其中，E 是能量密度，C_m 是根据充放电曲线计算的比电容，ΔV 是在放电时间为 Δt 时的电压变化，图 4.19（d）归纳了不同功率密度下电极材料的能量密度，N-pC-1073 在功率密度为 200 W·kg^{-1} 时的能量密度达到 19.0 Wh·kg^{-1}，显著高于 N-pC-973 和 N-pC-1173（分别为 17.0 Wh·kg^{-1} 和 12.1 Wh·kg^{-1}），这个值也比大部分的商用碳基超级电容器要高出不少[164]。N-pC 材料高的能量密度可以归结为碳骨架中掺杂的氮原子，它不仅可以增加材料的导电性，也能增加材料在水相电解液中的亲水性，从而增强了离子和电子的传输效率[129]。此外 N-pC 材料的多孔性结构也有利于电极材料和电解液表面的离子和电子传输,从而增加材料的功率和能量密度[165]。

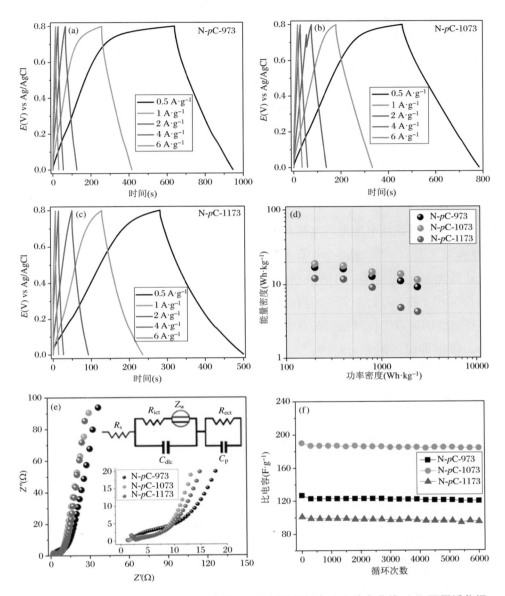

图 4.19 (a~c) 不同活化温度下制备的 N-pC 材料的恒电流充放电曲线;(d) 不同活化温度下制备的 N-pC 材料的能量比较图;(e) 不同活化温度下制备的 N-pC 材料的电化学阻抗谱;(f) 不同活化温度下制备的 N-pC 材料的长期循环稳定性评价

电化学阻抗谱用于进一步研究离子/电荷的传输和扩散动力学,以及材料的内阻。图 4.19(e)所示的是材料的电化学阻抗谱以及相应的等效电路[166]。在等效电路中,R_s 是参比电极(Ag/AgCl)和工作电极之间的溶液电阻,R_{ict} 是电极和电解液界面电荷传输的电阻,Z_w 是有限长度 Warburg 扩散阻抗,主要包括电解液离子从电极材料主体扩散至电极-电解液表面遇到的阻抗,R_{ect} 是氧化还原反应中电子传递过程的电阻，C_{dlc}是电双层电容，C_p是赝电容。这些组件构成了一个混合控制的 Randles 电路，经常用于模拟包含电双层电容和赝电容的超级电容器电极材料的电荷传输和扩散过程[167]。在低频区，所有 N-pC 材料的阻抗谱都呈现出一个大角度的直线，表明 N-pC 材料的结构有利于电子的传输和电解液离子的渗透[107]。而在高频区的阻抗谱发现一个小的弧形，可以解释为在电极材料和电解液界面之间存在高的离子导电性，类似的现象在单壁碳纳米管和还原石墨烯氧化物电容也有报道[167-168]。根据 N-pC-973，N-pC-1073 和 N-pC-1173 材料的阻抗谱计算出材料的内阻分别为 1.36 Ω，1.28 Ω 和 1.94 Ω，结果表明 N-pC-1073 电极材料具有最低的内阻和最好的导电性，这对其优异的电容性能至关重要。

作为一个超级电容器的电极材料，一个最重要的因素是其循环稳定性和耐用性。图 4.19(f)所示的是 N-pC 材料在 50 mV·s^{-1} 的扫描速率下，根据 CV 曲线计算出来的比电容，结果表明，经过 6000 次的 CV 扫描之后，N-pC-973，N-pC-1073 和 N-pC-1173 的电容损失分别为 5%，3% 和 6%。N-pC 材料的循环稳定性明显高于文献报道的其他氮掺杂多孔碳材料（通常电容损失率在扫描 2000 圈时即可达到 7% 以上）[169]。

在本节研究中，我们通过快速热解-KOH 活化的方法合成了氮掺杂的多孔碳材料，并将其用作超级电容器的电极材料。根据实验结果，可得出以下结论：

（1）快速热解过程可以将超过 50% 的蒲草生物质中的氮元素富集在热解生物炭中，而通过 KOH 的活化，可以大大增加材料的比表面积和改善孔隙结构。

（2）活化温度对材料的结构和氮含量影响很大，高的活化温度有利于材料孔隙的长大和比表面积的增加，但是会显著降低材料的氮含量，因此需要综合考虑材料的氮含量和比表面积等来优化活化温度。

（3）合成的氮掺杂多孔碳材料具有良好的电容性能，其比电容最大能达到 257 F·g^{-1}，并且具有优异的循环稳定性，可以循环 6000 次而电容下降很少。氮元素的掺杂可以增加材料的导电性和亲水性，对材料电容性能的提升起到至关重要的作用。

4.4

泡沫生物炭电极的制备及其电催化氧化 5-HMF

4.4.1

概述

层状双金属氢氧化物（Layered Double Hydroxide，LDH）又被称为水滑石，是一类由两种或两种以上金属元素组成的具有特定结构的混合金属氢氧化物。其结构由氢氧化物主层板和层间的插层阴离子及水分子相互交叠构成。由过渡金属组成的 LDH 往往具有一定的催化潜力，如用于电催化水产氧反应等[170]。LDH 本身结构细小，一般需要负载在导电基材上使用[171]。5-HMF 的电催化氧化反应是一个以传质扩散为主要阻力的反应，因此适合在快速搅拌的体系下进行。传统的碳纸、碳毡等材料的机械强度和亲水性较低，多次使用后容易发生破裂和催化剂的脱落。如何以生物炭为原料低成本地合成具有良好机械性能和稳定性的电极基材，是我们研究的重点。

甲壳素是世界上储量第二多的天然高分子，仅次于纤维素，广泛存在于虾壳等甲壳类动物的外骨骼中[172]。由于甲壳素不易加工，一般将甲壳素通过脱乙酰化处理为壳聚糖后再使用[173]。但是壳聚糖为线性高分子，其聚合物分子结构非常松散，若将其作为生物炭电极基材，则需要进一步加固其结构[174-175]。酚醛树脂是一类具有网状结构的高分子聚合物，具有较好的机械强度和热稳定性。因此我们尝试通过雷索酚-甲醛树脂（Resorcinol-Formaldehyde resin，RF 树脂，酚醛树脂的一种）加固壳聚糖材料。在 RF 树脂聚合过程中，甲醛分子还同时能桥接上壳聚糖分子中的氨基，与 RF 树脂的分子形成更加稳定的网状交联聚合物，最终构建出具有良好机械强度的壳聚糖材料[176]。

本节中，我们通过"冻干-热固-热解"三步法得到具有稳定结构和机械强度的泡沫生物炭，并以其作为基底负载 LDH 催化剂，构建具有 5-HMF 选择性氧化能力的催化电极。我们还运用一系列表征手段，深入研究了电极材料的理化性质。我们首先通过 SEM 和 EDS 对电极表面 LDH 层的形貌和元素组成进行了分析，再通过 XRD、XPS、拉曼光谱和红外吸收光谱进一步分析了电极表面 LDH 催化剂的物相组成。同时我们还利用拉曼光谱和接触角仪对生物炭电极基底的性质进行一定的分析。通过这些表征结果，我们可以深入了解生物炭电极基底和 LDH 催化剂的物化性质，这为分析其结合能力和催化机理提供理论支持。

接下来我们采用所制备的生物炭基底负载 LDH 催化电极对水氧化和 5-HMF 选择性氧化进行了实验。我们首先测试了电极的一系列电化学性能，并在不同反应条件下对 5-HMF 氧化进行了一系列实验。同时我们还利用核磁共振谱（Nuclear Magnetic Resonance，NMR）和数学模型对 5-HMF 的氧化过程和副反应机理进行了进一步解析。之后我们还成功分离了 FDCA 粗产物结晶，并通过 HPLC，NMR 等表征手段对产物的纯度进行了一系列的分析。最后我们还对电极的循环耐用性和法拉第效率等进行了分析，并提出相应的解决方法。

4.4.2
电极材料的表征结果

我们通过扫描电子显微镜分析了 LDH@CF 电极表面的形貌结构。为了维持电极表面原有的形貌，我们直接将小块电极放置在 SEM 载物台上进行观测。如图 4.20 所示，NiFe 层状双金属氢氧化物纳米片纵向密集生长在碳基底表面上，并形成致密的 LDH 层。LDH 层随着 CF 的粗糙表面弯曲生长并形成褶皱。通过图 4.21（a）的 SEM-EDS mapping 照片可以看到 LDH 层上 Ni，Fe 和 O 元素都是均匀分布的，而在 4.21（b）的照片上则可以观察到暴露的碳基底上几乎没有 Fe 元素的分布。同时在碳基底上分布有一定含量的 K 元素，而在 LDH 层上则几乎没有 K 元素分布。我们在碳基底和 LDH 催化剂的合成中并没有引入 K 元素，因此我们认为这些 K 元素是在 KOH 电解液中进行电化学实验时被碳基底捕获的 K^+ 离子。

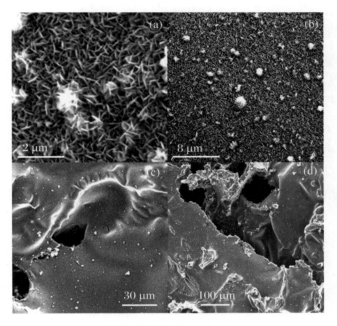

图 4.20 LDH@CF 的 SEM 照片

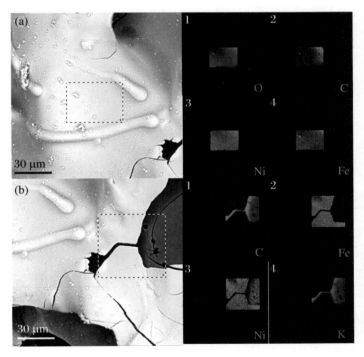

图 4.21 (a) LDH@CF 的 SEM-EDS mapping 照片(O,C,Fe,Ni);(b) 暴露出碳基底的 LDH@CF 的 SEM-EDS mapping 照片(C,Ni,Fe,K)

我们通过一系列的测试对 LDH@CF 电极材料进行了表征。图 4.22(a) 是 LDH@CF 的粉末 X 射线衍射谱,可以看到图谱中有两个属于无定形碳的宽峰。同时由于无定形碳宽峰的掩盖,只能明显观察到 NiFe LDH 的 3 个主要衍射峰中 12°左右的(003)晶面的衍射峰,(006)和(012)晶面的衍射峰则较难观察到。我们还通过傅里叶变换红外光谱仪对 LDH@CF 材料中的官能团进行了分析,可以观察到羟基的振动吸收峰,以及 CO_3^{2-}、H_2O、OCN^- 等水热合成法合成的 LDH 中常见的插层阴离子和水分子,证明了碳基底上负载的氢氧化物的确是具有插层结构的 LDH。图 4.22(c 和 d)显示了合成的 NiFe LDH 催化剂中 Ni 具有 Ni-OH 和 Ni-O 两种价态,而 Fe 则具有 Fe^{2+} 和 Fe^{3+} 两种物相,这些都是 NiFe LDH 中常见的形貌。

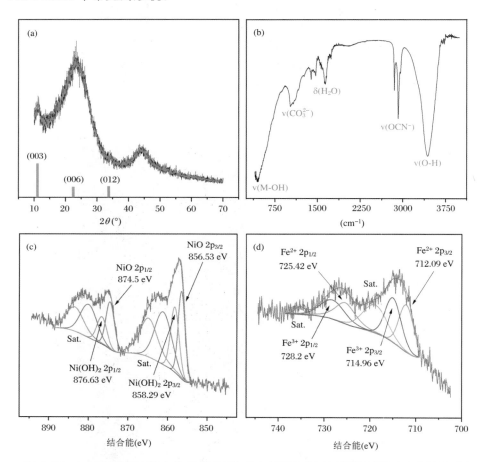

图 4.22 (a) LDH@CF 粉末的 XRD 图谱;(b) LDH@CF 的 FTIR 图谱;(c) LDH@CF XPS 图谱的 Ni 2p 分峰;(d) LDH@CF XPS 图谱的 Fe 2p 分峰

在对碳基底粉末进行 SEM 拍摄时我们注意到在粉末表面呈现的贝壳状裂纹,而拉曼光谱和 XRD 表征则显示该碳基底材料为高度无序化的无定形碳,该材料还具有和石墨相似的高机械强度和导电性。经过多项表征对比,我们认为该碳基底表面的材料为玻璃态碳[177]。如图4.23(c和d)所示,我们测试了玻璃态碳和普通碳纸的水接触角,玻璃态碳表现出远高于碳纸的亲水性。这可能是碳基底表面LDH纳米片纵向生长的原因。

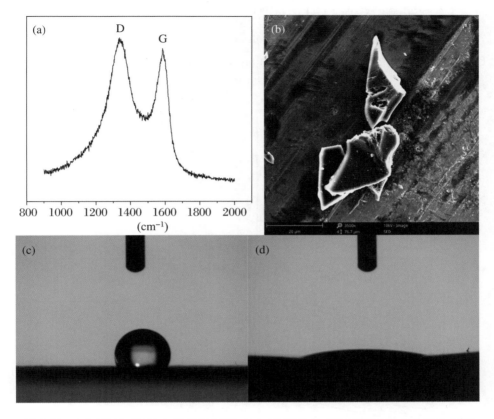

图4.23 (a) LDH@CF粉末的拉曼图谱;(b) CF粉末的SEM照片(放大倍数为3500倍);(c) 碳纸的水接触角测试照片;(d) CB的水接触角测试照片

4.4.3

生物炭电极的电化学性能

我们分别在 0.1 mol·L^{-1} KOH 溶液和 20 mmol·L^{-1} 5-HMF + 0.1 mol·L^{-1} KOH 溶液中对 NiFe LDH@CF 电极和未负载 LDH 的碳电极基底进行电化学测试。如图 4.24(a)所示,我们通过 LSV 曲线测试了 LDH@CF 电极的氧化过电势。LDH@CF 电极在 0.1 mol·L^{-1} KOH 溶液中的水氧化电位为 1.43 V,5-HMF 的氧化电位为 1.42 V,远高于未负载 LDH 催化剂的碳基底的氧化电位

(高于 1.62 V)。这说明 LDH 催化剂能够有效降低反应氧化过电势,促进水和 5-HMF 的氧化。图 4.24(b) 显示在 Tafel 斜率区,LDH 催化剂能够有效提高碳电极基底的氧化性能。通过在非法拉第区的 CV 循环测试显示,碳电极基底具有较大的比表面电容,而 LDH 的负载反而降低了电极的比表面电容。图 4.24(f) 的交流阻抗谱测试也显示 LDH 的负载反而增大了电极的传质电阻,这说明 LDH 的催化氧化能力并不是通过提高电极的比表面电容活性实现的。

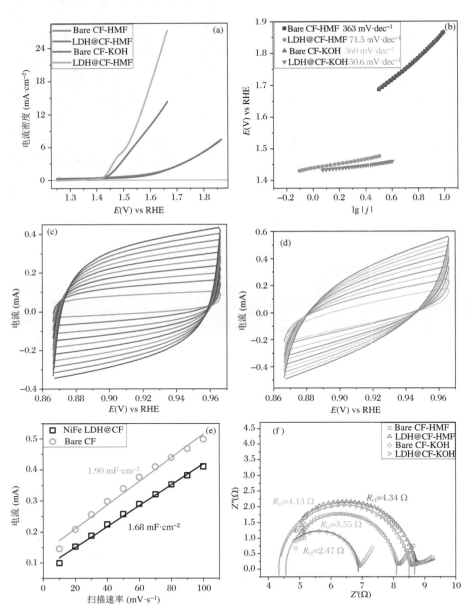

图 4.24 (a) LDH@CF 电极的 LSV 曲线(扫描速率为 1 mV·s^{-1});(b) LDH@CF 电极的 Tafel 斜率;(c) CF 电极的非法拉第区的 CV 曲线;(d) LDH@CF 电极的非法拉第区的 CV 曲线;(e) LDH@CF 电极的比表面电容;(f) LDH@CF 电极的交流阻抗谱曲线

同时我们还注意到,在工作电位足够发生 5-HMF/H_2O 氧化反应之前,LDH@CF 电极的 LSV 曲线并不是接近 0,而是从低电位开始就有一个较大的电流增长(图 4.25)。这表现为电容的充电效果。我们知道壳聚糖分子中存在着氨基,从而导致热解得到的是氮掺杂的碳材料。有关文献曾报道过氮掺杂的碳材料可以作为超级电容器填料[178],而之前的 SEM-EDS 照片也显示在碳基底中存在着大量的 K^+。这证明氮掺杂的碳基底材料在启动氧化反应之前大量捕获了溶液中的 K^+,从而产生了较大的充电电流。

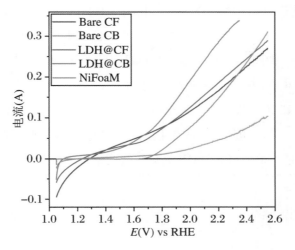

图 4.25 不同电极在 10 m mol·L^{-1} 5-HMF/0.1 mol·L^{-1} KOH 溶液中的 LSV 曲线

扫速为 1 mV·s^{-1}

我们合成了多种组分的 LDH 催化剂,并测试其对 5-HMF 的催化性能。可以看到在不同的 LDH 催化剂中,NiFe LDH 催化剂的 5-HMF 转化率和 FDCA 选择性都具有最好的效果(表 4.8)。因此后续的测试中,我们均选用 NiFe LDH@CF 催化电极进行实验。

表 4.8 不同 LDH 催化剂的性能

LDH@CF 电极	FDCA 选择率	5-HMF 转化率
NiFe	62.2%	75.9%
NiCo	2.56%	21.6%
NiCu	41.1%	65.1%
NiAl	28.9%	65.3%
CuAl	7.43%	37.9%
CoAl	15.9%	50.3%

我们在 1.41~1.71 V(vs RHE)的恒反应电位下进行 5-HMF 的氧化实验,反应的产物和中间物都通过 HPLC 进行检测。HPLC 使用 BioRad HPX-87H 硅胶柱作为分离柱,流动相为 5 mmol·L^{-1} 的硫酸且采用 0.6 mL·min^{-1} 流速的等度洗脱方式。在反应过程中每次移取 1 mL 样液并加入 11 mL 稀盐酸稀释中和电解液中的 KOH,取 10 μL 稀释后并用 0.22 μm 的滤头过滤的样液注入 HPLC 系统中。5-HMF,FDCA 等反应物和产物的出峰时间和浓度通过外标法计算得出。

随着反应的进行,5-HMF 逐渐被氧化为 5-HMFCA,DFF,FFCA 等氧化中间产物以及最终氧化产物 FDCA。我们在液相色谱中都检测到了这些物质,如图 4.26(a)所示。如图 4.26(c)所示,NiFe LDH@CF 催化电极能将 5-HMF 高效地转化为 FDCA 同时保持较高的选择性。一般情况下,10 mmol·L^{-1} 5-HMF 在反应 12 h 后基本消耗完全并转化为 FDCA,同时其他的中间产物一直保持在低浓度。而未负载 LDH 催化剂的 CF 电极基底则不具有这样的选择性催化功能,只是简单地将 5-HMF 氧化分解而未生成相关产物(图 4.26(b))。

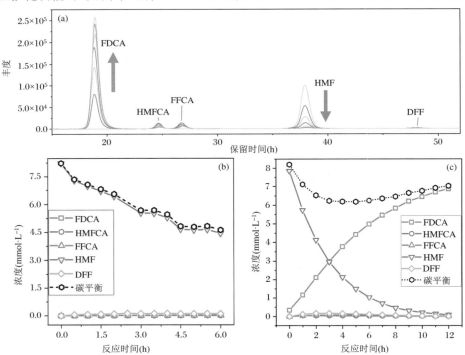

图 4.26　(a) 5-HMF 氧化过程中生成的产物和中间产物的液相色谱;(b) CF 电极氧化 5-HMF 时各产物的浓度变化(10 mmol·L^{-1});(c) LDH@CF 电极氧化 5-HMF 时各产物的浓度变化(10 mmol·L^{-1} 5-HMF);(d) LDH@CF 电极氧化 5-HMF 时各产物的浓度变化(20 mmol·L^{-1} 5-HMF);(e) LDH@CF 电极氧化 5-HMF 时各产物的浓度变化(20 mmol·L^{-1} 5-HMF,大电流)

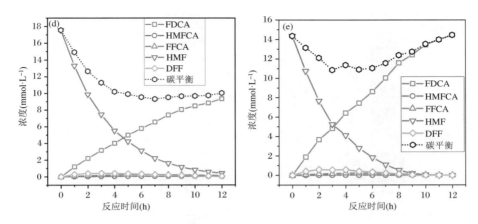

续图 4.26　(a) 5-HMF 氧化过程中生成的产物和中间产物的液相色谱;(b) CF 电极氧化 5-HMF 时各产物的浓度变化(10 mmol·L^{-1});(c) LDH@CF 电极氧化 5-HMF 时各产物的浓度变化(10 mmol·L^{-1} 5-HMF);(d) LDH@CF 电极氧化 5-HMF 时各产物的浓度变化(20 mmol·L^{-1} 5-HMF);(e) LDH@CF 电极氧化 5-HMF 时各产物的浓度变化(20 mmol·L^{-1} 5-HMF,大电流)

但是我们注意到,在整个反应过程中出现了碳平衡(反应前后有机物的总和)不守恒的情况。不同于一般耗散过程中碳平衡逐渐减少的现象,在我们的反应中碳平衡经历了一个先减少后上升的过程。在有些情况下碳平衡甚至能恢复到最初的水平,而在更高 5-HMF 浓度的情况下碳平衡的损失则更大(图 4.26(d))。碳平衡的变化表现为有机物离开了反应体系又逐渐回归的过程,这种情况我们一开始认为是碳电极基底对 5-HMF 的吸附作用导致的。于是我们进行了一组吸脱附实验验证这个假设:将未通电的 LDH@CF 电极浸泡在 10 mmol·L^{-1} 5-HMF/0.1 mol·L^{-1} KOH 溶液中 3 h,并检测吸附前后溶液中的 5-HMF 的浓度变化。如图 4.27(a)所示,吸附后溶液中的 5-HMF 浓度有所下降但是并没有如前反应中减少的那么多。这说明反应中出现的碳平衡不守恒现象另有原因。

5-HMF 在强碱性条件和空气中不稳定,会逐渐变质生成深色的化合物。而在 5-HMF 氧化的过程中我们注意到溶液的颜色先由浅变深再由深变浅,这说明 5-HMF 在反应过程中可能转化成了一些其他的化合物,且这个转化过程是可逆的(或是转化后的化合物也能够被氧化为 FDCA)。为了验证这一猜想,我们使用 5-HMF 氧化的中间产物 5-HMFCA 和 DFF 作为反应底物,验证其氧化过程中的碳平衡变化情况。如图 4.27(b 和 c)所示,5-HMFCA 在氧化过程中碳平衡基本保持稳定,而 DFF 则出现了明显的碳平衡下降。说明这种转化不是 5-HMF 独有的,DFF 等物质也能发生类似的现象。通过查阅相关文献我们了

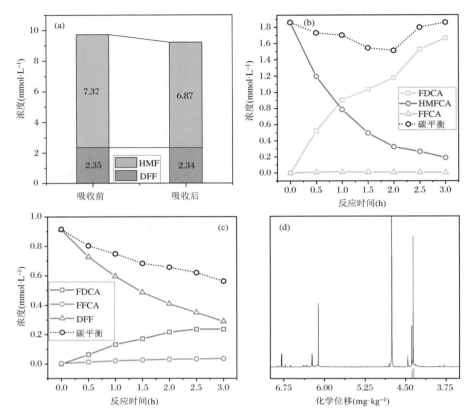

图 4.27 (a) 吸附前后溶液中 5-HMF/DFF 的浓度变化；(b) 5-HMFCA 作为 LDH@CF 电极氧化底物时各产物的浓度变化（2 mmol·L^{-1}）；(c) DFF 作为 LDH@CF 电极氧化底物时各产物的浓度变化（1 mmol·L^{-1}）；(d) 陈化后的 5-HMF 溶液的 ^{1}H-NMR 信号

解到，5-HMF 在一定条件下会发生聚合生成二聚物等低聚物[179]。这些低聚物的聚合过程是部分可逆的，随着时间的推移，这些聚合物将逐渐解聚或直接氧化为 FDCA，或进一步聚合变质为腐殖质。这种聚合作用依赖于分子上的醛基，因此没有醛基的 5-HMFCA 在氧化过程中则没有出现明显的碳平衡不守恒的现象。

由于这些低聚物在液相色谱上的出峰时间与 5-HMF 非常相近且具有较小的紫外吸收强度，通过外标法计算 5-HMF 浓度时即表现为 5-HMF 浓度偏低和碳平衡下降。为了验证低聚物的存在，我们通过 NMR 对体系中的 5-HMF 进行检测。我们在 1 mol·L^{-1} 氘氧化钠（NaOD）的重水溶液中加入了 100 mmol·L^{-1} 的 5-HMF，然后在静置 24 h 后用 NMR 检测已经变质为深红色的 5-HMF 溶液。如图 4.27(d) 所示，在 ^{1}H-NMR 图谱中可以检测到 5-HMF 低聚物的信号，说明陈化变质后的 5-HMF 溶液中出现了 5-HMF 的低聚物。

尽管这些 5-HMF 低聚物的聚合反应是可逆的，但是其他副产物如开环化合物和腐殖质等则不可能重新变为 5-HMF 或氧化为 FDCA。为了提高 FDCA 的产率，除了控制电解液 pH 外，还需要提高反应的速度以减少副反应发生的时间。因此我们通过提高反应电流的方法来加快 5-HMF 的氧化速度。提高反应电流主要通过两个方式实现：① 增大电极基底的尺寸；② 提高反应电势。如图 4.26(e) 所示，在提高反应电流以后，在高浓度的 5-HMF 条件下依然可以保持较高的 5-HMF 转化率和 FDCA 选择性，并完全恢复体系的碳平衡。但是高电位作用下体系也容易发生水分解的副反应，从而降低体系的法拉第效率。在极端情况下，体系的法拉第效率甚至低于 50%，即使在改善反应条件后体系的法拉第效率仍然只有 75%。但是考虑到高电流是提高反应速度减少 5-HMF 副反应产生重要方法，约 75% 的法拉第效率也是可以接受的。

体系的法拉第效率计算如下：

$$FE = \frac{\sum_t (C_t - C_{0t}) \cdot V \cdot N_t F}{\int_0^T A(t) \mathrm{d}t} \times 100\% \qquad (4.25)$$

其中，C_{0t} 是产物或中间物的初始浓度（在反应物不纯时不严格为 0）；C_t 是它们反应后的浓度；N_t 是 5-HMF 转化为该化合物时转移的电子数（例如 $N_{\mathrm{FDCA}} = 6$）；V 是电解液的体积；F 是法拉第常数。反应转移的电荷量则由 i-t 曲线积分得到（如图 4.28 所示）。

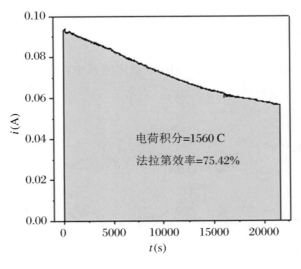

图 4.28　优化后反应体系的 i-t 曲线及其电荷积分和法拉第效率

如图 4.29 所示，5-HMF 氧化为 FDCA 存在两条反应路径。通过液相色谱我们在反应体系中均检测到了 DFF 和 5-HMFCA，说明两种反应路径在我们的体系中均存在。那么哪种反应路径是我们的 NiFe LDH@CF 催化氧化 5-HMF 的主要方式呢？[180]，我们假设反应过程中每一步单元反应都是一级动力学反应，以此建立反应动力学模型（如方程(4.26)所示）。虽然通过 5 个方程可以得到 5 个未知量的方程解，但由于 $D(t)$ 和 $E(t)$ 的表达式过于复杂，且我们关心的主要是 k_1 和 k_2 的相对大小。因此我们通过以 DFF 和 5-HMFCA 为反应底物的氧化实验测定了 k_3 和 k_4 的数值（图 4.30(a)），并代入方程(4.27)即可求得 k_1 和 k_2 的相对大小（其中 $M = k_1 + k_2$，可由原始数据直接得到）。如图 4.30(b) 所示，$k_{DFF} \gg k_{5\text{-}HMFCA}$，这说明在本反应体系中大多数 5-HMF 是通过 DFF 路径氧化至 FDCA 的。

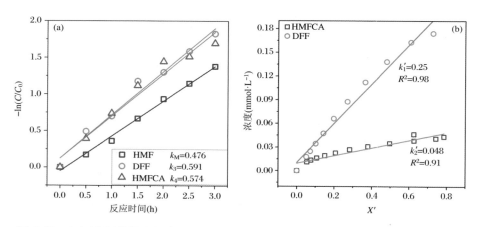

图 4.30 (a) 以 5-HMF，DFF 和 5-HMFCA 作为底物时反应的一级动力学常数；(b) 通过公式对 k_1' 和 k_2' 计算结果的线性拟合

至此，我们成功描述了在本 5-HMF 氧化体系中可能发生的反应及副反应过程，以及氧化的主要路径（图 4.31）。

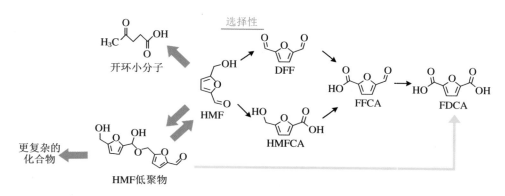

图 4.31 本 5-HMF 氧化体系中发生的所有反应

$$\begin{cases} \dfrac{dA}{dt} = -k_1 A - k_2 A \\ \dfrac{dB}{dt} = k_1 A - k_3 B \\ \dfrac{dC}{dt} = k_2 A - k_4 C \\ \dfrac{dD}{dt} = k_3 B + k_4 C - k_5 D \\ \dfrac{dE}{dt} = k_5 D \end{cases} \Rightarrow \begin{cases} A(t) = C_1 e^{t[-(k_1+k_2)]} \\ B(t) = C_2 e^{-k_3 t} - \dfrac{C_1 k_1 e^{-k_3 t}[e^{-t(k_1+k_2-k_3)} - 1]}{k_1+k_2-k_3} \\ C(t) = C_3 e^{-k_4 t} - \dfrac{C_1 k_2 e^{-k_4 t}[e^{-t(k_1+k_2-k_4)} - 1]}{k_1+k_2-k_4} \end{cases}$$

(4.26)

$$\Rightarrow \begin{cases} A(t) = A_0 e^{-Mt} \\ B(t) = A_0 \dfrac{k_1}{k_3 - M}(e^{-Mt} - e^{-k_3 t}) \\ C(t) = A_0 \dfrac{k_2}{k_4 - M}(e^{-Mt} - e^{-k_4 t}) \end{cases} \quad (4.27)$$

通过蒸发重结晶得到的 FCDA 粗产物粉末呈淡黄色(图 4.32)。纯净的 FDCA 粉末应为白色,而我们合成的 FDCA 粉末因吸附了少量 5-HMF 的有色副产物从而呈淡黄色。

图 4.32 分离得到的 FDCA 粗产物

为了检验粗产物 FDCA 的纯度,我们首先将 FDCA 粗产物溶于去离子水,并用 HPLC 分析其中的成分。由图 4.33 可以看到在产物中存在极微量的 5-HMFCA,FFCA 以及一些未知杂质(保留时间<20 min)。但是这些杂质的含量极少,由积分峰面积计算得到的 FDCA 纯度在 99.7% 以上。同时我们也对 FDCA 粗产物进行了一些表征,如 UV-vis,FTIR 和 NMR 等,并与相关文献中的 FDCA 标准品的图谱[181]进了比对(图 4.34)。这些表征结果尤其是 ^{13}C-NMR 谱都表明我们所合成的 FDCA 产物具有非常高的纯度。

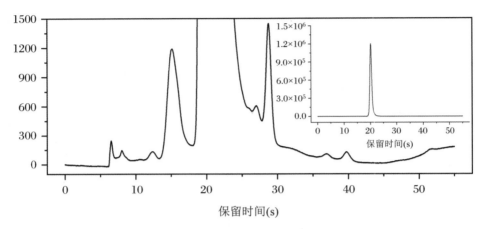

图 4.33　粗产物 FDCA 溶液的液相图谱的局部放大图

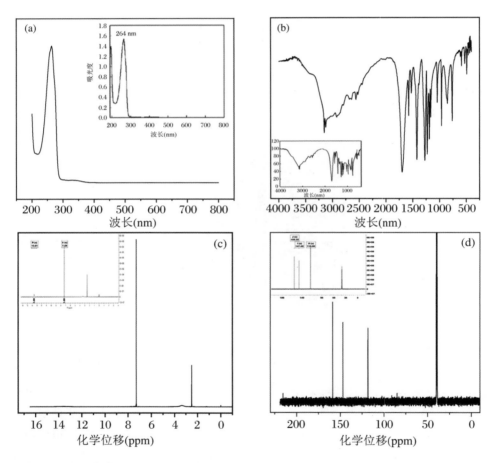

图 4.34　(a) 粗产物 FDCA(和 FDCA 标准品)的 UV-vis 吸收光谱；(b) 粗产物 FDCA(和 FDCA 标准品)的 FTIR 吸收光谱；(c) 粗产物 FDCA(和 FDCA 标准品)的 ^{1}H-NMR 信号；(d) 粗产物 FDCA(和 FDCA 标准品)的 ^{13}C-NMR 信号

为了测试电极材料的稳定性,我们进行了 5 轮循环实验(图 4.35)。虽然 LDH@CF 催化电极在 5 次循环实验后 5-HMF 的转化率和 FDCA 氧化选择性均保持在较高的水平。然而在循环实验的 $i\text{-}t$ 图中我们看到每次循环后电流都有非常大的衰减,这说明了电极导电能力的下降(尽管其催化能力没有明显的变化)。

同时我们对比了反应前后 LDH@CF 的 XPS 图谱。如图 4.36 所示,LDH@CF 的 Ni, Fe 元素在反应后的元素价态没有发生明显的变化,但是 C, O 的 XPS 图谱中则在反应后出现了明显的碳氧键的峰。这说明碳基底表面在反应过程中发生了一定程度的氧化,从而使得电极导电能力发生了下降。但是电极表面 LDH 层的性质没有发生改变,从而使其对 5-HMF 的选择性氧化能力没有发生变化。

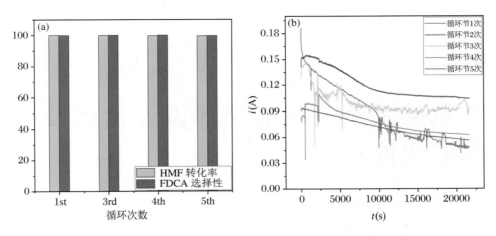

图 4.35 (a) 5 次循环实验的 5-HMF 转化率和 FDCA 选择性;(b) 5 次循环实验的 $i\text{-}t$ 曲线

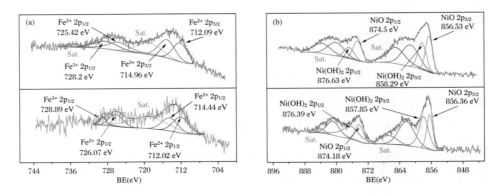

图 4.36 反应前后的 LDH@CF XPS 图谱的 Fe 2p/Ni 2p/C 1s/O 1s 分峰

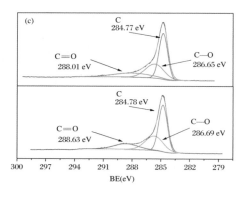
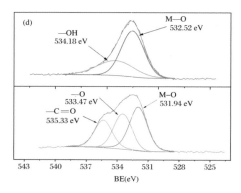

续图 4.36 反应前后的 LDH@CF XPS 图谱的 Fe 2p/Ni 2p/C 1s/O 1s 分峰

4.5 铁氮共掺杂生物炭材料设计合成及其电催化性能

4.5.1 概述

随着科技的发展和医疗水平的提升,人们健康保障系数不断提高。然而,药物的大量使用,甚至是抗生素的滥用,造成的危害仍然巨大。一方面,抗生素的滥用产生细菌耐药性会严重危害人类的健康;另一方面,抗生素作为排放到环境的污染物,对环境、生态以及人类生活同样构成了严重威胁。环境中的抗生素主要有以下几大来源:① 制药厂废水的排放;② 人类和动物体内抗生素,通过尿液和粪便代谢形式排放到环境中;③ 医院、居住地等人类生活场所排放的废水。各个来源排放的抗生素,在通过自然界流动的水环境作用下,对自然界的水体和土壤产生严重污染,进而从食物链循环影响着人类。因此,开发优异的抗生素废水处理工艺是当前所需要发展的项目。

此外,难降解有机污染物相关的水污染问题日益严重,近年来同样引起了全球的关注。为了解决这些问题[182-184],人们开发了各种基于活性氧(如羟基自由基($\cdot$OH)和超氧自由基($\cdot O_2^-$)等)形式的高级氧化过程(AOPs)。其中,以$Fe(Ⅲ)/Fe(Ⅱ)$循环活化过氧化氢生成$\cdot$OH为基础的芬顿工艺是应用较广泛的AOPs之一。该工艺操作简单且污染物降解效率高[185-186]。然而,传统芬顿工艺通常存在以下几个问题,比如:① pH适应性太窄(2.5~3.5);② 使用H_2O_2存在较高的安全风险;③ 含铁污泥产量大[187]。为了克服这些问题,文献中报道了对传统芬顿工艺的若干改进,其中非均相电芬顿是较为理想的一种,具有以下的优点[188]:① 电芬顿法可原位产生H_2O_2,避免了H_2O_2的外源添加及由此带来的安全风险;② 在电芬顿过程中,$Fe(Ⅱ)$和$\cdot$OH以相对恒定的速率连续产生,保证了较长的时间和有效的污染物降解,降低了含铁污泥的产量;③ 电芬顿法的pH适应性(2.0~6.5)比传统的芬顿法要宽得多。鉴于这些优点,电芬顿法在处理难降解的含有有机污染物废水方面起着重要的作用,近年来受到了广泛的关注[189-190]。

非均相电芬顿污染物降解效率很大程度上依赖于阴极催化剂,它不仅通过双电子转移O_2还原反应(ORR)来决定H_2O_2的生成,而且通过H_2O_2的活化来决定$\cdot$OH的生成[188,191-192]。在现有的先进阴极材料中,碳基材料,包括碳气凝胶、碳纳米管、活性炭是研究最广泛的电芬顿阴极材料[193-195]。由于其良好的导电性、易调控的孔隙结构和对H_2O_2的生成具有氧气还原双电子转移的高选择性,被认为是极有前途的电芬顿阴极材料。

作为一种高效的阴极电芬顿催化剂,催化H_2O_2活化生成$\cdot$OH的活性也至关重要。在碳材料上负载活性铁是制备高效阴极电芬顿催化剂的一种可行途径,该方法不仅可以对H_2O_2的ORR产生高的催化活性,而且为H_2O_2的活化提供了丰富的活性位点[196-199]。然而,基于碳材料电芬顿阴极催化剂的传统合成方法,通常涉及不可再生化石资源(碳材料的前驱体)、有害的或者爆炸性的化学物质(活性铁负载的配体或者还原剂)以及复杂的合成过程[200-201],这对实现环境友好和有效益的成本目标来说是困难的。

与传统的基于化石资源的碳材料相比,基于生物炭的材料具有多个优势[202-203]。首先,生物炭是由天然丰富的生物质生产的,可再生且成本低廉;其次,生物炭通常具有丰富的表面官能团和易于调控的孔结构,从而更有利于随后的活性铁物质的负载和结构调节。考虑到这些优点,采用生物炭用作电芬顿阴极的催化剂,可能会成为高级氧化工艺的理想平台。此外,为了提高生物炭材料对 ORR 以及 H_2O_2 活化的催化作用,将杂原子 N 和 Fe 掺杂到生物炭材料中,并且在该过程中,通过离子液体辅助生物质碳化过程,完成 Fe 和 N 元素掺杂进入到生物炭材料中。

作为非均相催化剂,再循环性应该是其实际应用的关键问题。之前的研究人员已经证明,某些基于生物炭的材料在电芬顿过程中具有良好的再循环性[197,204]。但是尚无进一步的研究来阐明为什么这些材料可能具有良好的再循环性。阐明这种机理对于合理设计和合成具有高性能及耐久性的非均相电芬顿催化剂将是非常有益的。

在这项工作中,我们通过离子液体辅助生物质碳化过程,合成了 Fe,N 掺杂的生物炭材料(标记为 Fe-N/biochar-T,T 是碳化温度)。并将合成后的材料,用作非均相电芬顿阴极催化剂来降解磺胺嘧啶。该类抗生素,是废水中广泛存在的一种典型的难降解有机污染物。合成的 Fe,N 共掺杂的生物炭材料,通过促进产过氧化氢的双电子转移氧还原反应,显著地提高了电化学芬顿进程污染物的降解效率,并且过氧化氢随后活化生成活性氧(·OH 和 ·O_2^-)。阴极催化剂在电芬顿循环进程中显示出了卓越的可循环性,并且该进程中催化活性不仅没有下降反而不断增加。经过非原位的高分辨透射电子显微镜和电子顺磁共振分析发现,界面晶相转变引导的更多活性催化位点的暴露,是催化剂具有如此优异可循环性能的主要原因。这项工作将为高性能、高耐久电芬顿催化剂的合理设计和合成提供新的思路。

4.5.2
Fe-N/biochar 材料的结构特征

在该项工作中,我们首先通过离子液体[Bmin][FeCl$_4$]辅助的生物质碳化的方法(图 4.37)合成了 Fe,N 共掺杂的生物炭材料。在合成过程中,[Bmin][FeCl$_4$]不仅提供了离子热反应介质,还同时充当了 Fe 和 N 元素的添加源。通过简单的有机反应合成了[Bmin][FeCl$_4$],然后将其与生物质混合而进行离子热处理,从而获得了预碳化的生物质样品。之后,将预碳化的样品进行进一步碳化,以便获得 Fe-N/biochar 材料。在二次碳化过程中,生物质被分解而产生一些挥发性成分,如 H_2、CO、碳氢化合物(C_xH_y)和氧化碳氢化合物

($C_{x'}H_{y'}O_{z'}$)。这些挥发性的成分可以充当还原剂,使其在高温下与[Bmin][FeCl$_4$]中的Fe(Ⅲ)反应,生成低价的Fe物质,例如Fe_3O_4,Fe,Fe_3C等。另外,[Bmin][FeCl$_4$]离子液体也通过经历分解的过程,在残留的碳基质中保留部分的氮物质。

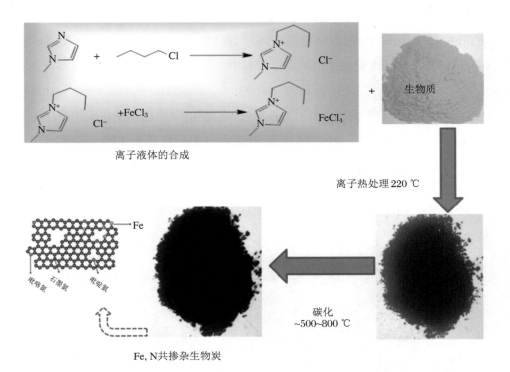

图4.37 合成Fe,N共掺杂生物炭材料的示意图

相比于浸渍或常规的化学合成方法,离子液体辅助生物质碳化方法具有很多优势[202]。首先,通过在生物碳化过程中就地热还原高阶铁,来实现活性铁物质的形成。这避免了使用额外的还原剂和配体。其次,生物炭和活性铁物质在同一过程中同时形成,易于操作并实现批量生产。再次,离子液体的组成易于调节,有利于将杂原子的掺杂定制为生物炭基质。最后,整个过程既环境友好又经济可持续。因为它既可以进行生物质处理,又可以进行资源回收以及材料的合成。据我们所知,这种实验操作方法还没有报道过。因此,这是第一次通过离子液体辅助生物质碳化过程制备多元素掺杂生物炭材料。

接下来,为进一步了解所得Fe-N/biochar材料的结构特征,我们使用了SEM,TEM和EDX,来分析Fe-N/biochar材料的组成和微观结构。Fe-N/biochar-600的SEM图像如图4.38(a)所示,图中展示了每个纽带是高度相互交联的,并形成了一个层次分明的碳网络。其中,包含了一些小颗粒附着在碳网络上。在 TEM 图像中,也进一步证实了小颗粒的存在(图 4.38(b))。在如图 4.38 中,HRTEM 的侧视图像显示了纳米粒子的高度结晶性。而且其连续的晶格条纹以及大约 0.203 nm 的晶面间距与金属 Fe 的(102)晶面很好的匹配。图 4.38(b2)边缘的连续晶格条纹可以归属于 Fe 氧化物的(006)晶面。材料颗粒的 SEAD 模式在图中呈现出,谱图 4.38(b3)中展示的分散的点,可以归结于金属铁。而谱图 4.38(b4)中展示的环形,可以归结于石墨碳壳。此外,Fe-N/biochar-600 的 EDS 元素谱图表明了材料中 Fe,N,O,C 共存。并且这些元素在所获得的材料中均匀分布(图 4.38(c)),这一结果也暗示了 N 和 Fe 被成功地引入生物炭的石墨碳基质中。为了进行比较,在图 4.39 中显示了在不同温度下制备的 Fe-N/biochar 材料的 SEM 图像,这表明高温可以改善生物炭碳基质中分层孔结构的形成。通过生物质直接碳化产生的原始生物炭的图像也显示在图 4.39(g~h)中。我们并没有在其中观察到该种样品分层的孔结构。从这些结果,我们可以推断出基于铁的离子液体极大地促进生物炭中分层孔结构的形成。

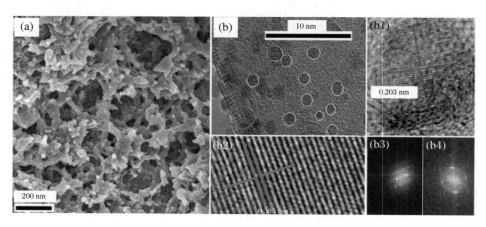

图 4.38 (a) Fe-N/biochar-600 样品的 SEM 图像;(b) Fe-N/biochar-600 样品的 TEM 图像;(b1) HRTEM 图像(0.203 nm);(b2) HRTEM 图像(0.335 nm)。(b3),(b4) Fe-N/biochar-600 样品的 SEAD 模式;(c) Fe-N/biochar-600 样品的 EDS 元素谱图;(d) Fe-N/biochar-600 样品的 XRD 谱图;(e) 氮气吸附-脱附等温曲线;(f) Fe-N/biochar-600 样品的 XPS C 1s 和(g) Fe-N/biochar-600 样品的 N 1s 光谱

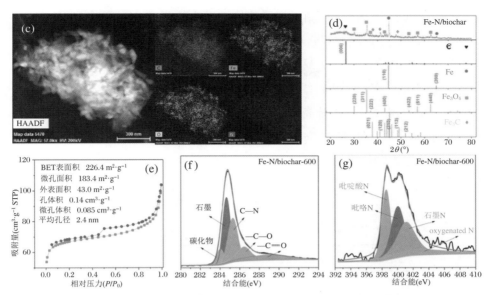

续图 4.38 (a) Fe-N/biochar-600 样品的 SEM 图像；(b) Fe-N/biochar-600 样品的 TEM 图像；(b1) HRTEM 图像(0.203 nm)；(b2) HRTEM 图像(0.335 nm)。(b3)，(b4) Fe-N/biochar-600 样品的 SEAD 模式；(c) Fe-N/biochar-600 样品的 EDS 元素谱图；(d) Fe-N/biochar-600 样品的 XRD 谱图；(e) 氮气吸附-脱附等温曲线；(f) Fe-N/biochar-600 样品的 XPS C 1s 和(g) Fe-N/biochar-600 样品的 N 1s 光谱

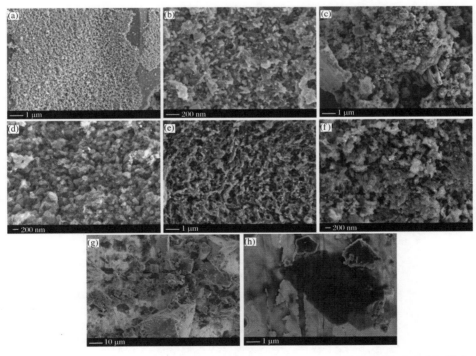

图 4.39 低倍率(a)和放大 SEM 图像(b) 的 Fe-N/biochar-500 样品；低倍率(c)和放大 SEM 图像(d) 的 Fe-N/biochar-700 样品；低倍率(e)和放大 SEM 图像(f) 的 Fe-N/biochar-800 样品；低倍率(g)和放大 SEM 图像(h) 的原始 biochar-600 样品

其次,我们采用 XRD 检测了离子液体辅助生物炭碳化过程中 Fe 的变化。如图 4.40 所示,当首次在 500 ℃下碳化生物质和离子液体时,在生成的生物炭材料中(Fe-N/biochar-500)只能发现 FeO 和 Fe_3O_4 物质。当温度升至 600 ℃时,FeO 物质消失了,而 X 射线衍射图像中却发现了两个新的物相,即金属 Fe 和 Fe_3C。随着温度进一步升高到 700 ℃和 800 ℃(图 4.40),并没有形成新的 Fe 物质,但是发现了更多的 Fe_3C 物质。XRD 的结果表明,随着温度的升高,Fe 的化学价态不断地降低,更多的 Fe 原子渗透到生物炭的碳基质中。

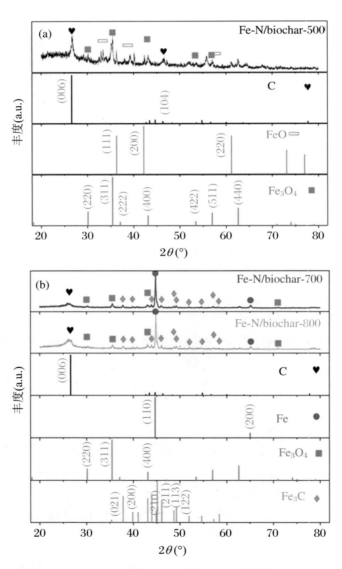

图 4.40　不同样品的 XRD 谱图:(a) Fe-N/biochar-500;
(b) Fe-N/biochar-700 和 Fe-N/biochar-800

对于使用离子液体辅助生物质碳化方法对生物炭孔结构产生的影响,采用了表征样品的氮气吸附/脱附等温线来考察。样品 Fe-N/biochar-600 的氮气吸附/脱附等温线呈现于图 4.38(e),根据 IUPA 分类显示,该图像属于Ⅳ类型的曲线[205]。该曲线表示了材料主要是具有微孔的柔性孔壁结构的多孔材料。根据该曲线计算得出,Fe-N/biochar-600 的 BET 表面积为 226.4 $m^2 \cdot g^{-1}$。比原始生物炭和 Fe-N/biochar-500 样品的 BET 表面积高得多,但比 Fe-N/biochar-700 和 Fe-N/biochar-800 样品要更低(图 4.41)。这一结果表明了基于铁的离子液体可以增强碳化过程中孔结构的形成,而且高温对孔的形成更加有利。

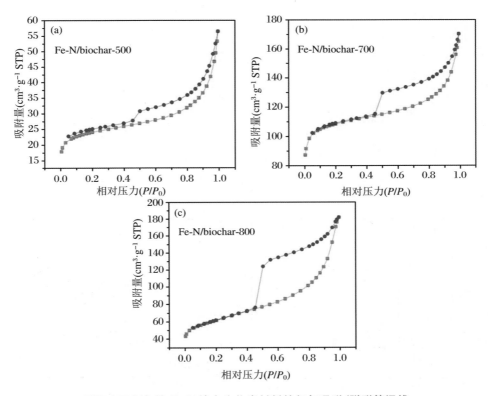

图 4.41　不同温度下制备的 Fe,N 掺杂生物炭材料的氮气吸附/脱附等温线

将 XPS 用于分析 Fe-N/biochar 材料的表面组成和化学价态。如图 4.42 所示,在不同温度下制备的 Fe-N/biochar 材料 XPS 全谱图表明,在不同温度下 Fe-N/biochar 材料所包含的元素大体一样,均含有 Fe,N,C,O 这四种元素。相比于其他温度下的材料,Fe-N/biochar-800 样品中未出现 Cl 元素的峰,可能是因为高温热处理的过程中,该元素形成了其他易挥发成分而挥发。此外,这些材料的表面 N 和 O 含量都随着温度的升高而连续降低。这种降低可能归因于高温下含 N 和 O 的物质(如 NH_3,HCN,CO,CO_2 等)释放的增加[202,206]。值得注意的是,表面铁含量也随着温度的升高而降低。这可以解释为随着温度的升高,生物质热解过程中形成的还原性挥发物质(例如 CH_4,C_2H_4,C_2H_2,H_2)可能沉

积在表面。铁物质表面形成碳纳米纤维或无定形碳（图 4.39）[207-208]，导致高温下获得的 Fe-N/biochar 样品表面 Fe 含量降低。

表 4.9　Fe-N/biochar 材料的表面积和孔隙结构

	Fe-N/biochar-500	Fe-N/biochar-600	Fe-N/biochar-700	Fe-N/biochar-800	Biochar-600
BET 表面积（$m^2 \cdot g^{-1}$）	82.2	226.4	366.3	211.9	25.6
微孔面积（$m^2 \cdot g^{-1}$）	52.7	183.4	288.9	87.5	25.8
外表面积（$m^2 \cdot g^{-1}$）	29.8	43.0	77.5	124.4	—
孔体积（$cm^3 \cdot g^{-1}$）	0.07	0.14	0.24	0.26	0.01
微孔体积（$cm^3 \cdot g^{-1}$）	0.024	0.085	0.134	0.040	0.012
平均孔径（nm）	3.5	2.4	2.6	5.0	1.4

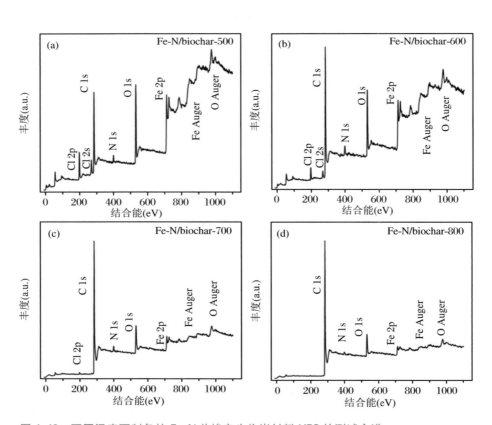

图 4.42　不同温度下制备的 Fe,N 共掺杂生物炭材料 XPS 的测试全谱

样品 Fe-N/biochar-600 的 XPS C 1s 谱图如图 4.38(f)所示，其可以拟合成结合能为 283.9 eV,284.6 eV,285.2 eV,286.2 eV 和 288.9 eV 的 5 个峰形，这些峰分别可以归属为碳化物、C—C、C—N、C—O 以及 C═O。为了进行比较，其他 Fe-N/biochar 材料和原始生物炭的 XPS C 1s 光谱如图 4.43 所示。从图中可以看出，金属碳化物在温度高的条件下，更易生成。这也与 XRD 测试晶体结构所得的结果相符合。样品 Fe-N/biochar-600 的 XPS N 1s 光谱（图 4.38(g)）可以拟合为 398.3 eV,400.5 eV,401.2 eV 和 403.2 eV 的 4 个峰形，这些峰分别可以归属为吡啶 N、吡咯 N、石墨 N 和氧化氮[209]。在 Fe-N/biochar-600 样品中，主要的 N 种类是吡啶 N 和吡咯 N，它们占据总 N 的 60% 以上。当碳化温度升至 800 ℃时，该数值急剧下降至不足 30%（图 4.44）。这一现象主要归因于以下事实：在高温碳化过程中，某些吡啶 N 和吡咯 N 通过催化剂中 Fe 类物质转化成了石墨 N。图 4.45 显示了不同温度下制备的 Fe-N/biochar 材料的 Fe 2p 光谱，通过去卷积处理观察到了其中的 10 个峰，并对应于 Fe 的碳化物、Fe^{2+}（Fe $2p_{3/2}$：710.72 eV,Fe $2p_{1/2}$：724.26 eV）、Fe^{3+}（Fe $2p_{3/2}$：712.03 eV,Fe $2p_{1/2}$：725.52 eV）以及其相对应的卫星峰。

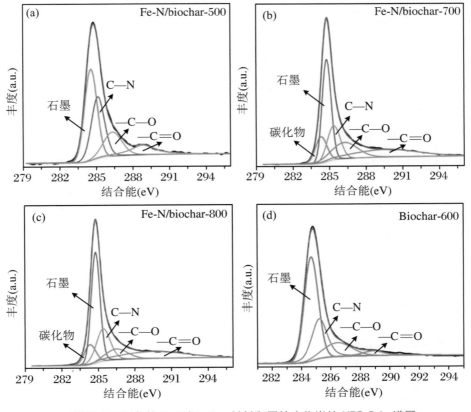

图 4.43　不同温度下制备的 Fe-N/biochar 材料和原始生物炭的 XPS C 1s 谱图

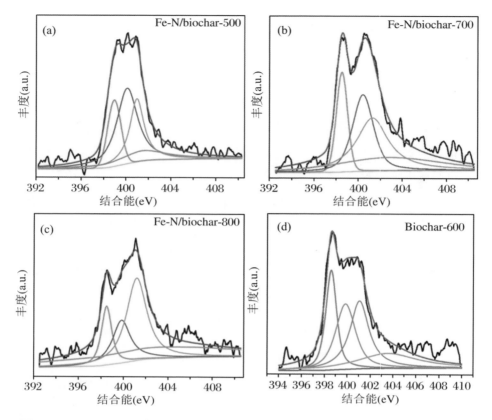

图 4.44 不同温度下制备的 Fe-N/biochar 材料和原始生物炭的 XPS N 1s 谱图

基于以上的结果,可以知道 Fe-N/biochar-600 材料表现出更高效的催化活性位点,即高比例的吡啶 N[210-211]、良好的电导率(由于其高的石墨化程度而引起)、大的表面积和丰富的互连微孔结构。所有这些因素,都对 Fe-N/biochar-600 材料具有高的电化学催化活性是有益的。

4.5.3
Fe-N/biochar 催化电芬顿降解磺胺嘧啶性能评价

为确保合成的材料与制备的三维电极适合进行电催化降解,我们首先进行了一些实验。图 4.46(a) 为电芬顿降解磺胺嘧啶的装置示意图,我们采用的是三电极体系来降解污染物。并且,首先通过测定不同材料三维电极的阻抗谱,来确保电极具有良好的导电性(图 4.46(b))。正如图中阻抗谱所呈现的,不同条件制备的生物炭材料都具有较小的阻抗,也即拥有良好的导电性。此外,从图中

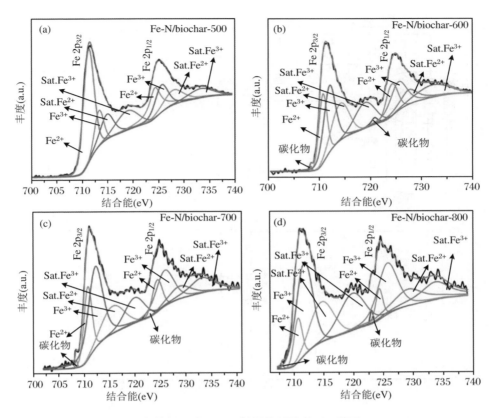

图 4.45　不同温度下制备的 Fe-N/biochar 材料的 XPS Fe 2p 谱图

还可以看到随着温度的升高,三维电极的阻抗谱逐渐减小。产生这一现象的原因可能亦如前所述,是由于温度的升高,生物炭石墨化程度提高,铁组分化学价态降低单质组分增加,从而增加了导电性。为了确定三维电极对磺胺嘧啶的吸附,以及材料电芬顿对溶液 pH 的影响,Fe-N/biochar-600 催化剂的相关实验呈现在如图 4.46(c 和 d)中。从图中可以看出,制备的 Fe-N/biochar-600 三维电极对溶液中磺胺嘧啶的吸附主要发生在前 20 min,且吸附较小。此外,相对于酸性溶液,Fe-N/biochar-600 材料电芬顿对碱性溶液的影响相对较大,但是均保持在不超过 1 的影响范围。所以,对于实验设计的体系,生物炭通过电芬顿降解抗生素磺胺嘧啶是适用的,不会受到其他条件过大的影响。

接下来,利用合成的 Fe-N/biochar 材料直接用作电芬顿体系中的阴极催化剂,来对抗生素磺胺嘧啶进行降解。如图 4.47 所示,Fe-N/biochar-600 催化剂在所有的 Fe,N 共掺杂生物炭材料中对磺胺嘧啶的降解表现出最佳性能,其速率常数为图 4.47 所呈现的 $0.024\ \mathrm{min}^{-1}$。相比之下,高于其他的催化剂的速率常数,如 Fe-N/biochar-500 为 $0.005\ \mathrm{min}^{-1}$,Fe-N/biochar-700 为 $0.006\ \mathrm{min}^{-1}$,Fe-N/biochar-800 为 $0.009\ \mathrm{min}^{-1}$。如表 4.10 所示,为了防止材料比表面积的

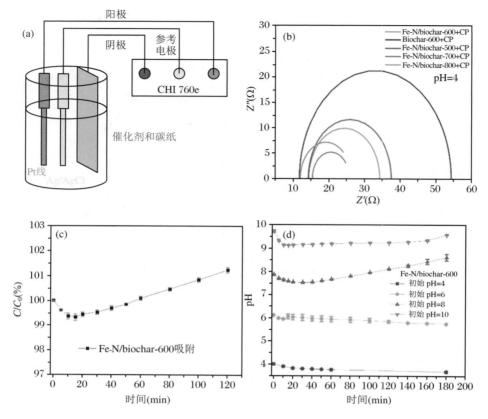

图 4.46　(a) 电催化降解磺胺嘧啶装置示意图;(b) 三维电极电化学阻抗谱; (c) Fe-N/biochar-600 材料对磺胺嘧啶的吸附;(d) 不同初始 pH 溶液在磺胺嘧啶降解过程中溶液 pH 的变化

影响,我们将其动力学速率常数进行了归一化。对 Fe-N/biochar-600 催化剂来说,其具有如此高的电芬顿活性与其独特的结构和表面组成密切相关。首先,据报道,吡啶 N 由于其丰富的孤对电子而普遍被认为是高效的氧还原催化活性位点[210-211]。因此,Fe-N/biochar-600 催化剂中吡啶 N 的高含量使其成为将 O_2 还原为 H_2O_2 的有利阴极催化剂。而且,Fe-N/biochar-600 催化剂的高表面 Fe 含量应该是影响其高电芬顿性能的另一个重要因素。因为在样品表面上高的 Fe 含量可以提供更多的催化位点来活化 H_2O_2,产生更多的 ·OH 自由基用于降解污染物。应该注意的是,尽管 Fe-N/biochar-500 样品比 Fe-N/biochar-600 样品和 Fe-N/biochar-800 样品具有更高的 Fe 与吡啶 N 含量,但是其表面积却要小得多,故而其电芬顿性能比那些材料要低得多。这一现象表明,在 N/Fe 含量和表面积之间可能需要权衡取舍,以增强阴极催化剂的电芬顿性能。而且,大的表面积可以创建更多的活性位点,以促进氧气的物理或化学吸附并改善其电化学还原成 H_2O_2 的能力[212-213]。

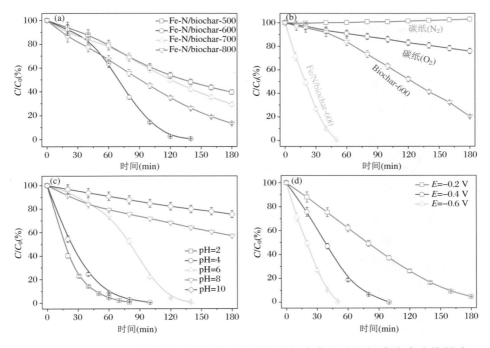

图 4.47　Fe-N/biochar 材料在不同条件下通过非均相电芬顿法降解磺胺嘧啶的影响：(a) 不同热解温度($pH=6.0, E=-0.4\,V$)；(b) 不同阴极催化剂和气流的影响($pH=4.0, E=-0.6\,V$)；(c) pH 的影响($E=-0.4\,V$)；(d) 阴极电位($pH=4.0$)

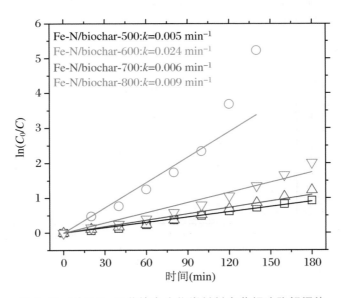

图 4.48　对于 Fe,N 共掺杂生物炭材料电芬顿法降解污染物动力学曲线

表 4.10　各个样品降解动力学常数对表面积的归一化

样品 Fe-N/biochar-X	500	600	700	800
k (min^{-1})	0.005	0.024	0.006	0.009
S_{BET} (m$^2 \cdot$ g^{-1})	82.2	226.4	366.3	211.9
k_s (mg $\cdot$ m$^{-2} \cdot$ min^{-1})	0.06	0.11	0.02	0.04

在图 4.49 中显示了一系列的对照实验,以确定电芬顿对磺胺嘧啶降解的贡献。结果表明,当我们使用原始碳纸作为阴极进行电化学测试时,在氮气流下 180 min 内几乎观察不到磺胺嘧啶降解(速率常数: $k = 0.00014$ min^{-1})。这表明了电吸附和电氧化对磺胺嘧啶的降解没有任何贡献。而在氧气流下,只有不到 20% 的磺胺嘧啶可以降解(速率常数: $k = 0.0015$ min^{-1})。这表明直接氧化对磺胺嘧啶的降解略有贡献。为了进行比较,原始生物炭同样也用作阴极催化剂,结果显示出 80% 的磺胺嘧啶可在 180 min 内降解(速率常数: $k = 0.0067$ min^{-1})。这一结果表明原始生物炭催化剂比原始碳纸对磺胺嘧啶的降解速度快得多,但是却比 Fe-N/biochar-600 催化剂的速度慢得多(速率常数: $k = 0.072$ min^{-1})。所以,基于上述实验,证实了电芬顿对污染物降解的主要作用。

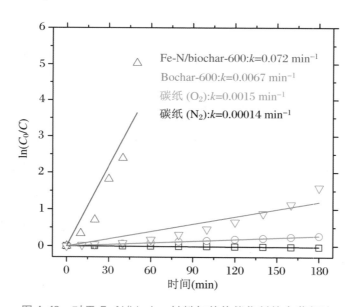

图 4.49　对于 Fe-N/biochar 材料与其他催化剂的电芬顿法降解磺胺嘧啶动力学曲线

图 4.47(c) 中显示了在不同 pH 下,磺胺嘧啶通过电芬顿的降解测试。实验表明,磺胺嘧啶在 pH = 2.0 下可于 80 min 内完全降解。并且当 pH 增加到 4.0

和 6.0 时，降解时间分别延长至 100 min 和 120 min。随着 pH 进一步增加到碱性范围内（pH 至 8.0 和 10.0），即使经过 180 min 的电芬顿降解后，仍然有超过一半的磺胺嘧啶残留。该实验表明，低的 pH 对电芬顿降解污染物工艺有很大的促进作用。据报道，pH 主要通过影响 Fe 物质的溶解度和 H_2O_2 的稳定性来影响电芬顿过程。首先，低的 pH 可以通过 2 电子转移氧还原反应（ORR）促进 H_2O_2 的产生和累计（图 4.47(c)）。并且，H_2O_2 可以被质子化形成稳定的氧离子[214]，从而可在 Fe(Ⅱ)物质的催化下进一步活化产生·OH 自由基。相反，随着 pH 的增加，由 ORR 新形成的 H_2O_2 可以直接分解成 H_2O 和 O_2，因此不会有 H_2O_2 的累积。此外，低的 pH 还可以促进催化剂中 Fe 的溶解，促进 H_2O_2 的活化以产生更多的·OH 自由基，从而提高电芬顿效率。因此，就此而言，在较低的 pH 下，电芬顿的效率更高。尽管 pH 为 2.0 是磺胺嘧啶降解的最佳 pH，但是较低的 pH 可能导致电芬顿反应器的快速腐蚀，并且由于铁物质的更快溶解而导致电催化剂的活性损失。因此，在实际应用中，pH 为 4.0 或者 6.0 对于可持续的电芬顿工艺应该更为可行。

电势对磺胺嘧啶降解的影响如图 4.47(d)所示，从图中我们可以看到，磺胺嘧啶完全降解的总时间从电势为 -0.6 V vs RHE 的 40 min 增加到 -0.2 V vs RHE 的 180 min 以上。在此条件下，磺胺嘧啶在更低负电位下降解的提升，可能是由 ORR 过程中产生更多的 H_2O_2 引起的[215-216]。这可以通过将 Fe-N/biochar 作为阴极催化剂，在 ORR 不同电势下的 CV 曲线得到证实（图 4.50）。然而，在电芬顿技术的实际应用中，应在所施加的阴极电势与降解效率之间进行权衡。因为，更低的负电势意味着更多的能源消耗。

为了研究不同催化剂通过电化学还原氧气产生过氧化氢的不同，我们分别测试了催化剂产生过氧化氢的选择性和累积产量。如图 4.51 所示，将催化剂通过 4 mol·L^{-1} 的盐酸酸洗 7 天，再测试其电还原氧气产生 H_2O_2 的能力。在使用二维环盘电极的测试过程中，不同材料在酸性（图 4.51(a)）与碱性（图 4.51(b)）条件下对 H_2O_2 的选择性完全不同。这说明了 Fe-N/biochar 样品在酸洗的过程中，内部可能包含 Fe_3C、单原子铁等易分解 H_2O_2 的成分。这些成分会在酸性溶液中无论是对 H_2O_2 选择性或者是累积产量都有一定的影响，因此所得数据与实际结果有一定的偏差。

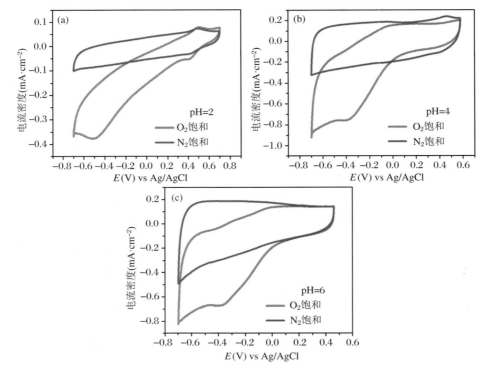

图 4.50 不同 pH 下 Fe-N/biochar-600 催化剂的电芬顿过程 CV 曲线图

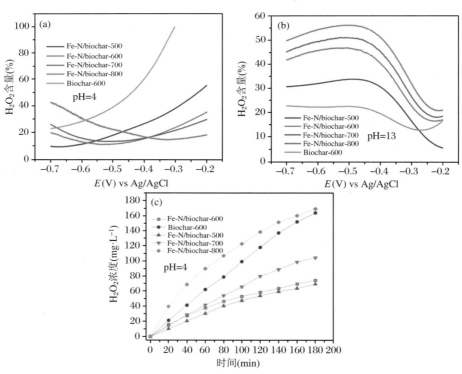

图 4.51 不同 Fe-N/biochar 材料通过环盘电极在 0.05 mol·L^{-1} Na$_2$SO$_4$ 溶液(a)、0.1 mol·L^{-1} KOH 溶液(b)中 H$_2$O$_2$ 的选择性和 H$_2$O$_2$ 累积产量(c)

4.5.4
Fe-N/biochar 催化机理及其循环稳定性

据报道,自由基是电芬顿过程中污染物降解的主要因素。因此,为了说明是哪种自由基作为该实验的主要因素,我们进行了一系列的自由基捕获实验。叔丁醇(TBA)和对苯醌(BQ)分别作为羟基自由基($\cdot$OH)和超氧自由基($\cdot O_2^-$)的清除试剂。在图4.52(a)中,显示了使用这些清除剂通过芬顿电极降解磺胺嘧啶的结果。实验表明,当$\cdot$OH被TBA捕获后,磺胺嘧啶的降解率仅为23%;而当$\cdot O_2^-$被BQ捕获后,磺胺嘧啶的降解率仅为34%;当$\cdot$OH和$\cdot O_2^-$同时被TBA和BQ分别捕获后,在180 min内几乎没有观测到磺胺嘧啶被降解。这意味着$\cdot$OH和$\cdot O_2^-$都是电芬顿过程中磺胺嘧啶降解的主要贡献。为了进一步证实这一说法,自由基捕获实验的EPR光谱如图4.52(b)所示,实验结果表明在电芬顿反应6 h后,依旧可以通过表征的方法同时检测到$\cdot$OH和$\cdot O_2^-$自由基。究其原因,是TBA和BQ分别对$\cdot$OH和$\cdot O_2^-$良好的捕获作用。

基于这些结果,我们可以提出在Fe-N/biochar样品催化的电芬顿过程中,磺胺嘧啶降解的可能机理。如图4.52(c)所示,在电芬顿体系的阴极中,首先在Fe-N/biochar催化剂的N掺杂碳基质上通过2电子转移过程将O_2还原为H_2O_2(反应1)。然后,原位生成的H_2O_2被Fe-N/biochar催化剂中的Fe(Ⅱ)物质活化,生成$\cdot$OH和$\cdot O_2^-$自由基(反应2和反应3)。并且,此过程中Fe(Ⅱ)本身转化为Fe(Ⅲ)。另外,催化剂还可以通过利用Fe(Ⅱ)直接激活O_2产生$\cdot O_2^-$自由基(反应4)。此后,原位产生的$\cdot$OH和$\cdot O_2^-$自由基进攻磺胺嘧啶分子,使得磺胺嘧啶分子逐渐降解并完全矿化为H_2O,CO_2和其他的一些无机物,例如N_2和NO_3^-。与此同时,Fe(Ⅱ)是通过在阴极的Fe(Ⅲ)经过单电子还原生成的(反应5),从而确保了电芬顿体系进行的连续性。

$$O_2 + 2H^+ + 2e^- \longrightarrow H_2O_2 \qquad (反应1)$$

$$Fe^{2+} + H_2O_2 + H^+ \longrightarrow Fe^{3+} + H_2O + \cdot OH \qquad (反应2)$$

$$Fe^{2+} + H_2O_2 \longrightarrow Fe^{3+} + 2H^+ + \cdot O_2^- + 2e^- \qquad (反应3)$$

$$Fe^{2+} + O_2 \longrightarrow Fe^{3+} + \cdot O_2^- \qquad (反应4)$$

$$Fe^{3+} + e^- \longrightarrow Fe^{2+} \qquad (反应5)$$

对于该电芬顿体系中产生的两种自由基,考虑到分辨哪种自由基在电芬顿催化中占主导地位,我们采用对苯二甲酸和XTT钠盐这两种探针试剂,来半定量地分别捕获$\cdot$OH和$\cdot O_2^-$自由基。如图4.53所示,分别测试了对苯二甲酸

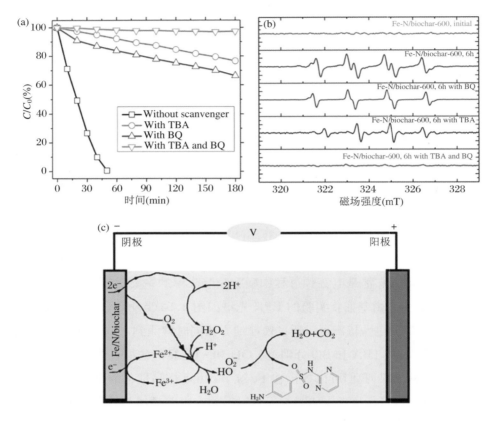

图 4.52 (a) 在存在不同自由基清除剂的情况下,通过电芬顿法降解磺胺嘧啶;(b) 在存在不同自由基清除剂的情况下电芬顿法的 ESR 光谱(捕获剂:BMPO);(c) 磺胺嘧啶电芬顿法的机理示意图

捕获·OH 自由基的荧光光谱强度,以及 XTT 钠盐捕获·O_2^- 自由基的紫外可见光谱强度。由于探针试剂具有专一性,该结果进一步的确定了 Fe-N/biochar 电芬顿体系中的自由基类型。此外,图 4.53(b)和图 4.53(d)中可以大体上看出荧光强度近似于线性增加,而紫外可见吸光度的增加速率递减。这暗示了 XTT 钠盐在电芬顿的过程中逐渐被·OH 氧化而对苯二甲酸却不易受·O_2^- 影响。结合上述淬灭实验和 ESR 捕获实验,因此推测在 Fe-N/biochar 电芬顿体系中,相比于·O_2^- 自由基,·OH 自由基的生成量可能占据主导地位。

为了获得有关磺胺嘧啶降解过程的更多信息,采用 LC-MS/MS 检测在电芬顿工艺过程中形成的可能中间体。表 4.11 列出了 24 种代表性中间体,根据它们的结构和浓度,我们可以推测通过电芬顿法降解磺胺嘧啶的可能途径(图 4.54)。由于官能团的空间效应,使磺胺嘧啶(化合物 A)首先在其 C1-N、C2 或者 C9 的位置被自由基(·OH 和·O_2^-)羟基化,形成羟胺(化合物 B)或酚衍生物(化合物 C 或 D)。然后水解羟基化的磺胺嘧啶使其中磺酰胺键断裂,从而形成几种单环化合物(化合物 E-I)。这些单环化合物可能会受到自由基的进一步攻击,导致更多的羟基附着在芳环上(化合物 J-N)。这些多羟基化合物不断地受到自由基的进攻,以解构芳环而形成数个开环化合物(化合物 O-R),并进一步降解为最终的矿化物质(H_2O,CO_2,N_2 和硫酸盐)。

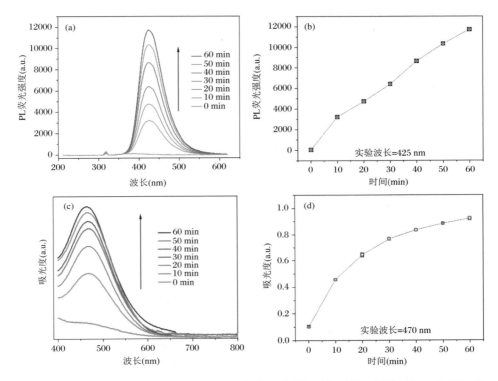

图 4.53 (a) 荧光强度随电芬顿进行增强全谱图(探针试剂:对苯二甲酸,$E=-0.6\,V$,pH $=5$,电解质 50 mmol·L^{-1} Na_2SO_4);(b) 波长 425 nm 处荧光强度随时间增长谱图;(c) 紫外可见光谱随电芬顿进行增强全谱图(探针试剂:XTT 钠盐,$E=-0.6\,V$,pH $=5$,电解质 50 mmol·L^{-1} Na_2SO_4);(d) 波长 470 nm 处紫外强度随时间增长谱图

表 4.11 HPLC-MS 分析的磺胺嘧啶电芬顿降解过程中潜在中间体的信息

化合物	分子式	m/z	可能的分子结构
A	$C_{10}H_{10}N_4O_2S$	250	

续表

化合物	分子式	m/z	可能的分子结构
B	$C_{10}H_{10}N_4O_3S$	266	4-(羟氨基)-N-(嘧啶-2-基)苯磺酰胺结构
C	$C_{10}H_{10}N_4O_3S$	266	4-氨基-N-(5-羟基嘧啶-2-基)苯磺酰胺结构
D	$C_{10}H_{10}N_4O_4S$	282	4-氨基-3-羟基-N-(5-羟基嘧啶-2-基)苯磺酰胺结构
E	$C_4H_5N_3$	95	2-氨基嘧啶结构
F	$C_6H_7NO_4S$	189	4-(羟氨基)苯磺酸结构
G	$C_6H_7NO_3S$	173	4-氨基苯磺酸结构

续表

化合物	分子式	m/z	可能的分子结构
H	$C_4H_5N_3O$	111	2-氨基-5-羟基嘧啶
I	$C_6H_7NO_4S$	189	4-氨基-3-羟基苯磺酸
K	$C_6H_7NO_5S$	205	4-(羟氨基)-3-羟基苯磺酸
L	C_6H_7N	93	苯胺
M	C_6H_7NO	109	4-氨基苯酚
N	C_6H_7NO	109	2-氨基苯酚
O	$C_2H_4O_2$	60	乙酸
P	$C_2H_2O_4$	90	草酸

续表

化合物	分子式	m/z	可能的分子结构
Q	$C_3H_4O_4$	104	
R	CH_5N_3	59	

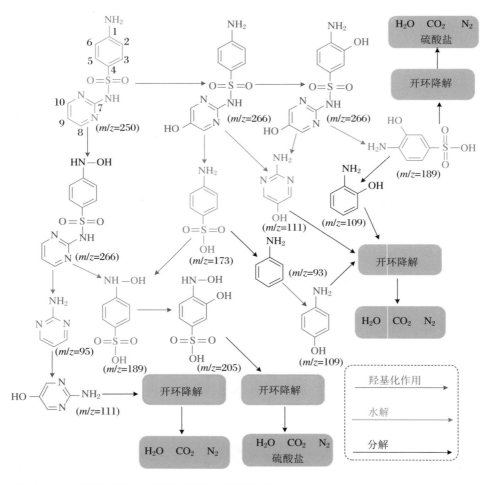

图 4.54　磺胺嘧啶通过电芬顿法可能的降解途径

作为一种非均相的电芬顿方法,阴极催化剂的可重复使用性是其应用潜力的关键因素。在实际废水处理中,如何高效、可持续地利用阴极催化剂是一个长期存在的问题。如图 4.55(a)所示,首次使用该催化剂时,电化学芬顿可以在 60 min 内降解磺胺嘧啶。而对于第一次的重复使用,磺胺嘧啶的降解时间降低为 40 min。在接下来的 10 次循环中,该数值下降到 30 min 甚至更少。这表明在 Fe-N/biochar 催化剂的电芬顿过程的循环运行中,其活性不断地增强。因此,计算了一个表观的周转频率(TOF)数值。其中,整个催化剂被视为催化的活性中心,因为我们无法计算出 Fe-N/biochar 的确切活性中心。如图 4.55(b) 所示,TOF 数值在电芬顿过程的 10 次循环中连续增加,进一步证实了在电芬顿过程的循环运行中 Fe-N/biochar 催化剂的性能得到了增强。

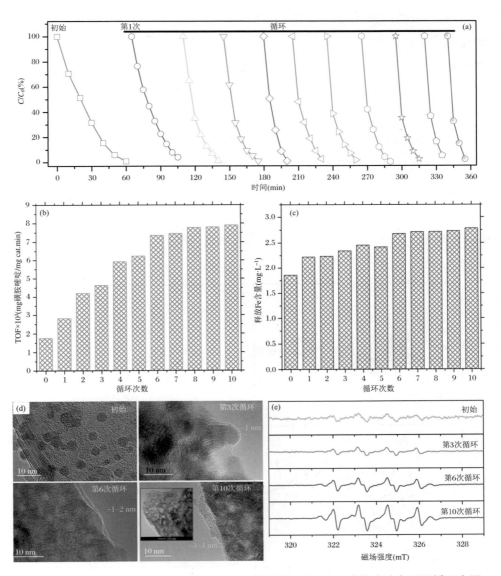

图 4.55 电芬顿过程中 Fe-N/biochar 催化剂的再循环性:(a) 磺胺嘧啶在不同循环中不同反应时间的剩余分数 (C/C_0);(b) 不同周期的表观 TOF 数值;(c) 在循环使用催化剂过程中 Fe 的释放含量;(d) 初始和循环后催化剂的 HRTEM;(e) 在不同电芬顿循环之后,阴极溶液的 EPR 光谱测试结果(捕获剂:BMPO)

令人好奇的是，Fe-N/biochar 的催化活性在电芬顿循环过程中并没有减弱，而是不断地增强，这与以往的报道有很大的不同。我们通过一系列探索性实验，分析了这一有趣现象的可能机制。首先，分析了每次循环运行过程中 Fe 的释放，在如图 4.55(c)的结果中，表明了在循环运行过程中 Fe 含量的释放持续增加。这与其 TOF 数值的增加相一致，表明 Fe 的释放对循环运行中 Fe-N/biochar 催化剂性能的增强起了重要的作用。

对循环性能起增强作用的重要贡献可能是在电芬顿过程中，Fe-N/biochar 催化剂界面晶相的转变。为了探测活性相是否随着电芬顿过程的进行而改变，在循环试验之前（标记为初始样品）和之后，对 Fe-N/biochar 催化剂进行了一系列的非原位 HRTEM 分析。如图 4.55(d)所示，在 Fe-N/biochar 催化剂的边缘发现了一层约 1 nm 的非常薄的层，这可能是由于催化剂边缘的金属 Fe 纳米颗粒形成的非晶 Fe(Ⅱ)O·Fe(Ⅲ)OOH 物质。在初始催化剂中几乎看不到该层的存在，并且其厚度（黄色虚线）随着循环的使用而增加。应该注意的是，这种晶体相变过程只发生在催化剂的边缘层，对催化剂本体材料的结构稳定性影响并不大。事实上，如图 4.56 所示，经过 10 次循环使用后，在 Fe-N/biochar 催化剂的 Fe 2p 和 C 1s 的 XPS 谱图中，并没有发现新的物质，从而证实了催化剂的稳定性。

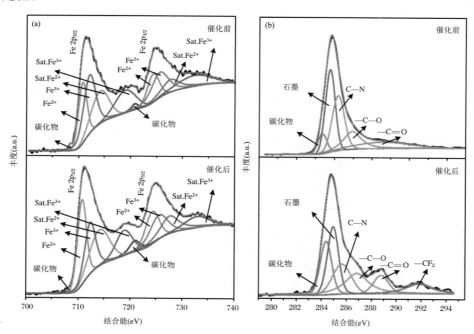

图 4.56　Fe-N/biochar-600 催化剂在 10 次电芬顿进程循环前后的 XPS Fe 2p 和 C 1s 光谱图

我们推测 Fe-N/biochar 催化剂在循环使用的过程中，其晶相转变可以暴露出更多的活性位点，从而可以提高阴极氧的还原能力，使体系中产生更多的 H_2O_2，并促使 H_2O_2 的活化得到更多的自由基，如 $\cdot OH$ 和 $\cdot O_2^-$，因此导致了磺胺嘧啶的快速降解。为了验证上述假设，我们首先对再循环的催化剂进行了阴极氧还原实验。但结果表明，再循环的催化剂氧还原性并未随着再循环次数的增加而增加。这意味着 H_2O_2 的产量，不会随着再循环次数的增加而增加。由于 $\cdot OH$ 自由基只能由 H_2O_2 形成，因此我们可以得出结论，在循环的电芬顿过程中，$\cdot OH$ 自由基并不是增强污染物降解性能的主要因素。与 $\cdot OH$ 自由基不同，$\cdot O_2^-$ 自由基可以通过一个电子转移的过程(反应 4 和图 4.52(c))，在含 Fe 化合物的催化下，由 O_2 的活化直接形成。因此，$\cdot O_2^-$ 自由基可能是污染物降解性能增强的原因之一。所以，我们进行了一系列的 EPR 测试以验证上述假设。如图 4.55(e)所示，随着循环电芬顿过程的进行，$\cdot O_2^-$ 自由基的 EPR 信号强度持续增强。这说明在催化剂再循环的过程中，$\cdot O_2^-$ 自由基的生成得到了提高。在循环电芬顿过程中，$\cdot O_2^-$ 自由基的生成是污染物降解性能改善的主要原因。

Fe-N/biochar 材料对磺胺嘧啶的电芬顿降解具有优异的催化活性和可回收性，这促使我们探索其对实际废水的处理性能。为此，我们利用 Fe-N/biochar 催化剂的电芬顿方法处理了一家抗生素工厂的实际废水。如表 4.12 所示，我们选择了其主要成分包含一种广谱性头孢类药物 7ACA 的废水，经过 9 h 的电芬顿处理，废水中超过 73% 的初始 COD 可以被去除(从 740.01 $mg \cdot L^{-1}$ 到 200.24 $mg \cdot L^{-1}$)，总有机碳去除率为 62%(图 4.57)。同时，废水中超过 93% 的磷和 23% 的氮也可以被去除。

表 4.12　使用 Fe-N/biochar-600 催化剂电芬顿降解实际废水前后的特性

TOC($mg \cdot L^{-1}$)	COD	NH_4^+ (N)	NO_2^- (N)	NO_3^- (N)	PO_4^{3-} (P)
降解前	740.01	43.76	0	0	6.96
降解后	200.24	33.51	0	0	0.48

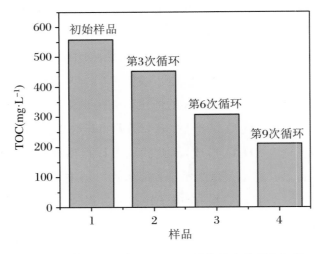

图 4.57 使用 Fe-N/biochar-600 催化剂电芬顿方法处理实际废水时 TOC 的变化情况

参考文献

[1] Solomon S, Plattner G-K, Knutti R, et al. Irreversible climate change due to carbon dioxide emissions [J]. Proceedings of the National Academy of Sciences, 2009, 106(6): 1704-1709.

[2] Woolf D, Amonette J E, Street-Perrott F A, et al. Sustainable biochar to mitigate global climate change [J]. Nat Commun, 2010, 1: 56.

[3] Broda M, Kierzkowska A M, Müller C R. Influence of the calcination and carbonation conditions on the CO_2 uptake of synthetic Ca-based CO_2 sorbents [J]. Environmental Science & Technology, 2012, 46(19): 10849-10856.

[4] Allen M R, Frame D J, Huntingford C, et al. Warming caused by cumulative carbon emissions towards the trillionth tonne [J]. Nature, 2009, 458(7242): 1163-1166.

[5] Wang S P, Yan S L, Ma X B, et al. Recent advances in capture of carbon dioxide using alkali-metal-based oxides [J]. Energy & Environmental Science, 2011, 4(10): 3805-3819.

[6] Jiménez V, Ramírez-Lucas A, Díaz J A, et al. CO_2 Capture in different carbon materials [J]. Environmental Science & Technology, 2012, 46(13): 7407-7414.

[7] Wang J T, Long D H, Zhou H H, et al. Surfactant promoted solid amine sorbents for CO_2 capture [J]. Energy & Environmental Science, 2012, 5(2): 5742-5749.

[8] Yasipourtehrani S, Tian S, Strezov V, et al. Development of robust CaO-based sorbents from blast furnace slag for calcium looping CO_2 capture [J]. Chemical Engineering Journal, 2020, 387: 124140.

[9] Yan X Y, Li Y J, Ma X T, et al. CeO_2-modified $CaO/Ca_{12}Al_{14}O_{33}$ bi-functional material for CO_2 capture and H_2 production in sorption-enhanced steam gasification of biomass [J]. Energy, 2020, 192: 116664.

[10] Wei S Y, Han R, Su Y L, et al. Size effect of calcium precursor and binder on CO_2 capture of composite CaO-based pellets [J]. Energy Procedia, 2019, 158: 5073-5078.

[11] Liu K, Zhao B, Wu Y, et al. Bubbling synthesis and high-temperature CO_2 adsorption performance of CaO-based adsorbents from carbide slag [J]. Fuel, 2020, 269: 117481.

[12] Hu Y, Lu H, Liu W, et al. Incorporation of CaO into inert supports for enhanced CO_2 capture: A review [J]. Chemical Engineering Journal, 2020, 396: 125253.

[13] Guo H, Xu Z, Jiang T, et al. The effect of incorporation Mg ions into the crystal lattice of CaO on the high temperature CO_2 capture [J]. Journal of CO_2 Utilization, 2020, 37: 335-345.

[14] Gao N B, Chen K L, Quan C. Development of CaO-based adsorbents loaded on charcoal for CO_2 capture at high temperature [J]. Fuel, 2020, 260: 116411.

[15] Chen J, Shi T, Duan L, et al. Microemulsion-derived, nanostructured CaO/CuO composites with controllable particle grain size to enhance cyclic CO_2 capture performance for combined Ca/Cu looping process [J]. Chemical Engineering Journal, 2020, 393: 124716.

[16] Abreu M, Teixeira P, Filipe R M, et al. Modeling the deactivation of CaO-based sorbents during multiple Ca-looping cycles for CO_2 post-combustion capture [J]. Computers & Chemical Engineering, 2020, 134: 106679.

[17] Valverde J M, Perejon A, Perez-Maqueda L A. Enhancement of fast CO_2 capture by a nano-SiO_2/CaO composite at Ca-looping conditions [J]. Environmental Science & Technology, 2012, 46(11): 6401-6408.

[18] Feng B, An H, Tan E. Screening of CO_2 Adsorbing materials for zero emission power generation systems [J]. Energy & Fuels, 2007, 21(2): 426-434.

[19] Bhagiyalakshmi M, Lee J Y, Jang H T. Synthesis of mesoporous magnesium oxide: Its application to CO_2 chemisorption [J]. International Journal of Greenhouse Gas Control, 2010, 4(1): 51-56.

[20] Meis N N A H, Bitter J H, De Jong K P. Support and size effects of activated hydrotalcites for precombustion CO_2 capture [J]. Industrial & Engineering Chemistry Research, 2009, 49(3): 1229-1235.

[21] Bian S-W, Baltrusaitis J, Galhotra P, et al. A template-free, thermal decomposition method to synthesize mesoporous MgO with a nanocrystalline framework and its application in carbon dioxide adsorption [J]. Journal of Materials Chemistry, 2010, 20(39): 8705-8710.

[22] Campelo J M, Luna D, Luque R, et al. Sustainable preparation of supported metal nanoparticles and their applications in catalysis [J]. ChemSusChem, 2009, 2(1): 18-45.

[23] De Villiers S, Dickson J A D, Ellam R M. The composition of the continental river weathering flux deduced from seawater Mg isotopes [J]. Chemical Geology, 2005, 216(1/2): 133-142.

[24] Huang Q Z, Lu G M, Wang J, et al. Thermal decomposition mechanisms of $MgCl_2 \cdot 6H_2O$ and $MgCl_2 \cdot H_2O$ [J]. Journal of Analytical and Applied Pyrolysis, 2011, 91(1): 159-164.

[25] Liu W J, Tian K, Jiang H, et al. Selectively improving the bio-oil quality by catalytic fast pyrolysis of heavy-metal-polluted biomass: Take copper (Cu) as an example [J]. Environmental Science & Technology, 2012, 46: 7849-7856.

[26] Liu W J, Zeng F X, Jiang H, et al. Techno-economic evaluation of the integrated biosorption-pyrolysis technology for lead (Pb) recovery from aqueous solution [J]. Bioresource Technology, 2011, 102(10): 6260-6265.

[27] Rouquerol J R, Sing K. Adsorption by powders and porous solids [M]. London: Academic Press, 1999.

[28] Valente J S, Tzompantzi F, Prince J, et al. Adsorption and photocatalytic degradation of phenol and 2,4 dichlorophenoxiacetic acid by Mg-Zn-Al layered double hydroxides [J]. Applied Catalysis B: Environmental, 2009,

90(3/4): 330-338.

[29] Dawson R, Stockel E, Holst J R, et al. Microporous organic polymers for carbon dioxide capture [J]. Energy & Environmental Science, 2011, 4(10): 4239-4245.

[30] Xiao G, Singh R, Chaffee A, et al. Advanced adsorbents based on MgO and K_2CO_3 for capture of CO_2 at elevated temperatures [J]. International Journal of Greenhouse Gas Control, 2011, 5(4): 634-639.

[31] Lee S C, Chae H J, Lee S J, et al. Development of regenerable MgO-based sorbent promoted with K_2CO_3 for CO_2 capture at low temperatures [J]. Environmental Science & Technology, 2008, 42(8): 2736-2741.

[32] Qi G, Wang Y, Estevez L, et al. High efficiency nanocomposite sorbents for CO_2 capture based on amine-functionalized mesoporous capsules [J]. Energy & Environmental Science, 2011, 4(2): 444-452.

[33] Sevilla M, Fuertes A B. Sustainable porous carbons with a superior performance for CO_2 capture [J]. Energy & Environmental Science, 2011, 4(5): 1765-1771.

[34] Mason J A, Sumida K, Herm Z R, et al. Evaluating metal-organic frameworks for post-combustion carbon dioxide capture via temperature swing adsorption [J]. Energy & Environmental Science, 2011, 4(8): 3030-3040.

[35] Wei H, Deng S, Hu B, et al. Granular bamboo-derived activated carbon for high CO_2 adsorption: The dominant role of narrow micropores [J]. ChemSusChem, 2012, 5(12): 2354-2360.

[36] Li L, Wen X, Fu X, et al. MgO/Al_2O_3 sorbent for CO_2 capture [J]. Energy & Fuels, 2010, 24(10): 5773-5780.

[37] Bhagiyalakshmi M, Hemalatha P, Ganesh M, et al. A direct synthesis of mesoporous carbon supported MgO sorbent for CO_2 capture [J]. Fuel, 2011, 90(4): 1662-1667.

[38] Martavaltzi C S, Lemonidou A A. Development of new CaO based sorbent materials for CO_2 removal at high temperature [J]. Microporous and Mesoporous Materials, 2008, 110(1): 119-127.

[39] Cui S, Cheng W W, Shen X D, et al. Mesoporous amine-modified SiO_2 aerogel: A potential CO_2 sorbent [J]. Energy & Environmental Science, 2011, 4(6): 2070-2074.

[40] Li L, King D L, Nie Z, et al. MgAl$_2$O$_4$ spinel-stabilized calcium oxide absorbents with improved durability for high-temperature CO$_2$ capture [J]. Energy & Fuels, 2010, 24(6): 3698-3703.

[41] Rufford T E, Hulicova-Jurcakova R, Zhu R, et al. A comparative study of chemical treatment by FeCl$_3$, MgCl$_2$, and ZnCl$_2$ on microstructure, surface chemistry, and double-layercapacitance of carbons from waste biomass[J]. Journal of Materials Research, 2010, 25(8):1451-1459.

[42] Morishita T, Soneda Y, Tsumura T, et al. Preparation of porous carbons from thermoplastic precursors and their performance for electric double layer capacitors [J]. Carbon, 2006, 44(12): 2360-2367.

[43] Morishita T, Tsumura T, Toyoda M, et al. A review of the control of pore structure in MgO-templated nanoporous carbons [J]. Carbon, 2010, 48(10): 2690-2707.

[44] Konno H, Onishi H, Yoshizawa N, et al. MgO-templated nitrogen-containing carbons derived from different organic compounds for capacitor electrodes [J]. Journal of Power Sources, 2010, 195(2): 667-673.

[45] Xing W, Liu C, Zhou Z Y, et al. Superior CO$_2$ uptake of N-doped activated carbon through hydrogen-bonding interaction [J]. Energy & Environmental Science, 2012, 5(6): 7323.

[46] Zhang Z J, Xu M Y, Wang H H, et al. Enhancement of CO$_2$ adsorption on high surface area activated carbon modified by N$_2$, H$_2$ and ammonia [J]. Chemical Engineering Journal, 2010, 160(2): 571-577.

[47] Zhou J H, Sui Z J, Zhu J, et al. Characterization of surface oxygen complexes on carbon nanofibers by TPD, XPS and FT-IR [J]. Carbon, 2007, 45(4): 785-796.

[48] Wakiya N, Kuroyanagi K, Xuan Y, et al. An XPS study of the nucleation and growth behavior of an epitaxial Pb(Zr,Ti)O$_3$/MgO(100) thin film prepared by MOCVD [J]. Thin Solid Films, 2000, 372(1/2): 156-162.

[49] Daub C D, Patey G N, Jack D B, et al. Monte Carlo simulations of the adsorption of CO$_2$ on the MgO(100) surface [J]. Journal of Chemical Physics, 2006, 124(11):2547.

[50] Jensen M B, Pettersson L G M, Swang O, et al. CO$_2$ sorption on MgO and CaO surfaces: A comparative quantum chemical cluster study [J]. The Journal of Physical Chemistry B, 2005, 109(35): 16774-16781.

[51] Miao S. Investigation on NIR, coating mechanism of PS-b-PAA coated calcium carbonate particulate [J]. Applied Surface Science, 2003, 220(1/2/3/4): 298-303.

[52] Zhou C H, Xia X, Lin C X, et al. Catalytic conversion of lignocellulosic biomass to fine chemicals and fuels [J]. Chemical Society Reviews, 2011, 40(11): 5588.

[53] Hara M. Biomass conversion by a solid acid catalyst [J]. Energy & Environmental Science, 2010, 3(5): 601.

[54] Taarning E, Osmundsen C M, Yang X, et al. Zeolite-catalyzed biomass conversion to fuels and chemicals [J]. Energy & Environmental Science, 2011, 4(3): 793.

[55] Tanabe K, Hölderich W F. Industrial application of solid acid-base catalysts [J]. Applied Catalysis A: General, 1999, 181(2): 399-434.

[56] Takagaki A, Tagusagawa C, Hayashi S, et al. Nanosheets as highly active solid acid catalysts for green chemical syntheses [J]. Energy & Environmental Science, 2010, 3(1): 82.

[57] Nakajima K, Hara M. Amorphous carbon with SO_3H groups as a solid brønsted acid catalyst [J]. ACS Catalysis, 2012, 2(7): 1296-1304.

[58] Xing R, Liu N, Liu Y, et al. Novel solid acid catalysts: sulfonic acid group-functionalized mesostructured polymers [J]. Advanced Functional Materials, 2007, 17(14): 2455-2461.

[59] Liu F J, Kong W P, Qi C Z, et al. Design and synthesis of mesoporous polymer-based solid acid catalysts with excellent hydrophobicity and extraordinary catalytic activity [J]. ACS Catalysis, 2012, 2(4): 565-572.

[60] Akiyama G, Matsuda R, Sato H, et al. Cellulose hydrolysis by a new porous coordination polymer decorated with sulfonic acid functional groups [J]. Advanced Materials, 2011, 23(29): 3294-3297.

[61] Tagusagawa C, Takagaki A, Iguchi A, et al. Highly active mesoporous Nb-W oxide solid-acid catalyst [J]. Angewandte Chemie International Edition, 2010, 49(6): 1128-1132.

[62] Li W, Jiang Z J, Ma F Y, et al. Design of mesoporous SO_4^{2-}/ZrO_2-SiO_2(Et) hybrid material as an efficient and reusable heterogeneous acid catalyst for biodiesel production [J]. Green Chemistry, 2010, 12(12): 2135.

[63] Gürbüz E I, Gallo J M R, Alonso D M, et al. Conversion of hemicellulose

into furfural using solid acid catalysts in γ-valerolactone [J]. Angewandte Chemie International Edition, 2013, 52(4):1270-1274.

[64] Hu X, Lievens C, Li C Z. Acid-catalyzed conversion of xylose in methanol-rich medium as part of biorefinery [J]. ChemSusChem, 2012, 5(8): 1427-1434.

[65] Lange J-P, Van De Graaf W D, Haan R J. Conversion of furfuryl alcohol into ethyl levulinate using solid acid catalysts [J]. ChemSusChem, 2009, 2(5): 437-441.

[66] Wang J J, Ren J W, Liu X H, et al. Direct conversion of carbohydrates to 5-hydroxymethylfurfural using Sn-mont catalyst [J]. Green Chemistry, 2012, 14(9): 2506-2512.

[67] Shuai L, Pan X J. Hydrolysis of cellulose by cellulase-mimetic solid catalyst [J]. Energy & Environmental Science, 2012, 5(5): 6889-6894.

[68] Wang S, Zhang L Q, Sima G B, et al. Efficient hydrolysis of bagasse cellulose to glucose by mesoporous carbon solid acid derived from industrial lignin [J]. Chemical Physics Letters, 2019, 736: 136808.

[69] Wang S, Sima G B, Cui Y, et al. Efficient hydrolysis of cellulose to glucose catalyzed by lignin-derived mesoporous carbon solid acid in water [J]. Chinese Journal of Chemical Engineering, 2020, 28(7):1866-1874.

[70] Tang X C, Niu S L. Preparation of carbon-based solid acid with large surface area to catalyze esterification for biodiesel production [J]. Journal of Industrial and Engineering Chemistry, 2019, 69: 187-195.

[71] Shi Y, Liang X. Novel carbon microtube based solid acid from pampas grass stick for biodiesel synthesis from waste oils [J]. Journal of Saudi Chemical Society, 2019, 23(5): 515-524.

[72] Shi Y, Kevin M-W, Liang X. Facile synthesis of novel porous carbon based solid acid with microtube structure and its catalytic activities for biodiesel synthesis [J]. Journal of Environmental Chemical Engineering, 2018, 6(5): 6633-6640.

[73] Li Y, Zeng D L. Synthesis and characterization of flower-like carbon spheres solid acid from glucose for esterification [J]. Materials Letters, 2017, 193: 172-175.

[74] Li M F, Zhang Q T, Luo B, et al. Lignin-based carbon solid acid catalyst prepared for selectively converting fructose to 5-hydroxymethylfurfural [J].

Industrial Crops and Products, 2020, 145: 111920.

[75] Hood Z D, Cheng Y, Evans S F, et al. Unraveling the structural properties and dynamics of sulfonated solid acid carbon catalysts with neutron vibrational spectroscopy [J]. Catalysis Today, 2020, 358: 387-393.

[76] Ballotin F C, Da Silva M J, Lago R M, et al. Solid acid catalysts based on sulfonated carbon nanostructures embedded in an amorphous matrix produced from bio-oil: Esterification of oleic acid with methanol [J]. Journal of Environmental Chemical Engineering, 2020, 8(2): 103674.

[77] Arancon R A, Barros Jr H R, Balu A M, et al. Valorisation of corncob residues to functionalised porous carbonaceous materials for the simultaneous esterification/transesterification of waste oils [J]. Green Chemistry, 2011, 13(11): 3162.

[78] Gill C S, Price B A, Jones C W. Sulfonic acid-functionalized silica-coated magnetic nanoparticle catalysts [J]. Journal of Catalysis, 2007, 251(1): 145-152.

[79] Lai D M, Deng L, Guo Q X, et al. Hydrolysis of biomass by magnetic solid acid [J]. Energy & Environmental Science, 2011, 4(9): 3552.

[80] Feyen M, Weidenthaler C, Schüth F, et al. Synthesis of structurally stable colloidal composites as magnetically recyclable acid catalysts [J]. Chemistry of Materials, 2010, 22(9): 2955-2961.

[81] Zillillah Z, Tan G W, Li Z. Highly active, stable, and recyclable magnetic nano-size solid acid catalysts: Efficient esterification of free fatty acid in grease to produce biodiesel [J]. Green Chemistry, 2012, 14(11): 3077-3086.

[82] Zafiropoulos N A, Ngo H L, Foglia T A, et al. Catalytic synthesis of biodiesel from high free fatty acid-containing feedstocks [J]. Chemical Communications, 2007(35): 3670-3672.

[83] Jones C W, Tsuji K, Davis M E. Organic-functionalized molecular sieves as shape-selective catalysts [J]. Nature, 1998, 393(6680): 52-54.

[84] Tagusagawa C, Takagaki A, Iguchi A, et al. Synthesis and characterization of mesoporous Ta-W oxides as strong solid acid catalysts [J]. Chemistry of Materials, 2010, 22(10): 3072-3078.

[85] Liu Z, Zhang F S. Nano-zerovalent iron contained porous carbons developed from waste biomass for the adsorption and dechlorination of PCBs [J]. Bioresource Technology, 2010, 101(7): 2562-2564.

[86] Maldonado-Hódar F J, Moreno-Castilla C, Rivera-Utrilla J, et al. Catalytic graphitization of carbon aerogels by transition metals [J]. Langmuir, 2000, 16(9): 4367-4373.

[87] Lim S F, Zheng Y M, Zou S W, et al. Characterization of copper adsorption onto an alginate encapsulated magnetic sorbent by a combined FT-IR, XPS, and mathematical modeling study [J]. Environmental Science & Technology, 2008, 42(7): 2551-2556.

[88] Toupin M, Belanger D. Spontaneous functionalization of carbon black by reaction with 4-nitrophenyldiazonium cations [J]. Langmuir, 2008, 24(5): 1910-1917.

[89] Liu F J, Sun J, Zhu L F, et al. Sulfated graphene as an efficient solid catalyst for acid-catalyzed liquid reactions [J]. Journal of Materials Chemistry, 2012, 22(12): 5495.

[90] Maciá-Agulló J A, Sevilla M, Diez M A, et al. Synthesis of carbon-based solid acid microspheres and their application to the production of biodiesel [J]. ChemSusChem, 2010, 3(12): 1352-1354.

[91] Li X T, Jiang Y J, Shuai L, et al. Sulfonated copolymers with SO_3H and COOH groups for the hydrolysis of polysaccharides [J]. Journal of Materials Chemistry, 2012, 22(4): 1283.

[92] Wu Y Y, Fu Z H, Yin D L, et al. Microwave-assisted hydrolysis of crystalline cellulose catalyzed by biomass char sulfonic acids [J]. Green Chemistry, 2010, 12(4): 696.

[93] Eberts T J, Sample R H, Glick M R, et al. A simplified, colorimetric micromethod for xylose in serum or urine, with phloroglucinol. [J]. Clinical Chemistry, 1979(8):1440-1443.

[94] Tong D S, Xia X, Luo X P, et al. Catalytic hydrolysis of cellulose to reducing sugar over acid-activated montmorillonite catalysts [J]. Applied Clay Science,2013,74:147-153.

[95] Lam E, Majid E, Leung A C W, et al. Synthesis of furfural from xylose by heterogeneous and reusable nafion catalysts [J]. ChemSusChem, 2011, 4(4): 535-541.

[96] Agirrezabal-Telleria I, Requies J, Guemez M B, et al. Furfural production from xylose + glucose feedings and simultaneous N_2-stripping [J]. Green Chemistry, 2012,14(11):3132.

[97] Simon P, Gogotsi Y. Materials for electrochemical capacitors [J]. Nature Materials, 2008, 7(11): 845-854.

[98] Aricò A S, Bruce P, Scrosati B, et al. Nanostructured materials for advanced energy conversion and storage devices [J]. Nature Materials, 2005, 4(5): 366-377.

[99] Zhang L L, Zhao X. Carbon-based materials as supercapacitor electrodes [J]. Chemical Society Reviews, 2009, 38(9): 2520-2531.

[100] Simon P, Gogotsi Y. Materials for electrochemical capacitors [J]. Nat Mater, 2008, 7(11): 845-854.

[101] Miller J R, Simon P. Electrochemical capacitors for energy management [J]. Science Magazine, 2008, 321(5889): 651-652.

[102] Chmiola J, Yushin G, Gogotsi Y, et al. Anomalous increase in carbon capacitance at pore sizes less than 1 nanometer [J]. Science, 2006, 313 (5794): 1760-1763.

[103] Zhao L, Fan L Z, Zhou M Q, et al. Nitrogen-containing hydrothermal carbons with superior performance in supercapacitors [J]. Advanced Materials, 2010, 22(45): 5202-5206.

[104] Kou Y, Xu Y H, Guo Z Q, et al. Supercapacitive energy storage and electric power supply using an aza-fused π-conjugated microporous framework [J]. Angewandte Chemie International Edition, 2011, 50(37): 8753-8757.

[105] Lu X H, Zheng D Z, Zhai T, et al. Facile synthesis of large-area manganese oxide nanorod arrays as a high-performance electrochemical supercapacitor [J]. Energy & Environmental Science, 2011, 4 (8): 2915-2921.

[106] Xie K, Qin X T, Wang X Z, et al. Carbon nanocages as supercapacitor electrode materials [J]. Advanced Materials, 2012, 24(3): 347-352.

[107] Wang L, Mu G, Tian C G, et al. Porous graphitic carbon nanosheets derived from cornstalk biomass for advanced supercapacitors [J]. ChemSusChem, 2013, 6(5): 880-889.

[108] Noked M, Okashy S, Zimrin T, et al. Composite carbon nanotube/carbon electrodes for electrical double-layer super capacitors [J]. Angewandte Chemie International Edition, 2012, 51(7): 1568-1571.

[109] Lota G, Fic K, Frackowiak E. Carbon nanotubes and their composites in

electrochemical applications [J]. Energy & Environmental Science, 2011, 4 (5): 1592-1605.

[110] Weng Z, Su Y, Wang D W, et al. Graphene-cellulose paper flexible supercapacitors [J]. Advanced Energy Materials, 2011, 1(5): 917-922.

[111] Wen Z H, Wang X C, Mao S, et al. Crumpled nitrogen-doped graphene nanosheets with ultrahigh pore volume for high-performance supercapacitor [J]. Advanced Materials, 2012, 24(41): 5610-5616.

[112] Zhang G L, Guan T T, Wang N, et al. Small mesopore engineering of pitch-based porous carbons toward enhanced supercapacitor performance [J]. Chemical Engineering Journal, 2020, 399: 125818.

[113] Sui Z Y, Chang Z S, Xu X F, et al. Direct growth of MnO_2 on highly porous nitrogen-doped carbon nanowires for asymmetric supercapacitors [J]. Diamond and Related Materials, 2020, 108: 107988.

[114] Shanmuga-Priya M, Divya P, Rajalakshmi R. A review status on characterization and electrochemical behaviour of biomass derived carbon materials for energy storage supercapacitors [J]. Sustainable Chemistry and Pharmacy, 2020, 16: 100243.

[115] Sathyamoorthi S, Tubtimkuna S, Sawangphruk M. Influence of structures and functional groups of carbon on working potentials of supercapacitors in neutral aqueous electrolyte: In situ differential electrochemical mass spectrometry [J]. Journal of Energy Storage, 2020, 29: 101379.

[116] Sathyamoorthi S, Tejangkura W, Sawangphruk M. Turning carbon-$ZnMn_2O_4$ powder in primary battery waste to be an effective active material for long cycling life supercapacitors: In situ gas analysis [J]. Waste Management, 2020, 109: 202-211.

[117] Liu H, Gao F Y, Fan Q C, et al. Preparation of novel 3D hierarchical porous carbon membrane as flexible free-standing electrode for supercapacitors [J]. Journal of Electroanalytical Chemistry, 2020: 114409.

[118] Li C, Zhang X, Wang K, et al. Recent advances in carbon nanostructures prepared from carbon dioxide for high-performance supercapacitors [J]. Journal of Energy Chemistry, 2021(3):16.

[119] Khawaja M K, Khanfar M F, Oghlenian T, et al. Fabrication and electrochemical characterization of graphene-oxide supercapacitor electrodes with activated carbon current collectors on graphite substrates

[J]. Computers & Electrical Engineering, 2020, 85: 106678.

[120] Gurten-Inal I I, Aktas Z. Enhancing the performance of activated carbon based scalable supercapacitors by heat treatment [J]. Applied Surface Science, 2020, 514: 145895.

[121] Gao Y, Zhang J, Luo X, et al. Energy density-enhancement mechanism and design principles for heteroatom-doped carbon supercapacitors [J]. Nano Energy, 2020, 72: 104666.

[122] Dong J X, Li S J, Ding Y. Anchoring nickel-cobalt sulfide nanoparticles on carbon aerogel derived from waste watermelon rind for high-performance asymmetric supercapacitors [J]. Journal of Alloys and Compounds, 2020, 845: 155701.

[123] Bordjiba T, Mohamedi M, Dao L H. New class of carbon-nanotube aerogel electrodes for electrochemical power sources [J]. Advanced Materials, 2008, 20(4): 815-819.

[124] Hulicova-Jurcakova D, Seredych M, Lu G Q, et al. Combined effect of nitrogen- and oxygen-containing functional groups of microporous activated carbon on its electrochemical performance in supercapacitors [J]. Advanced Functional Materials, 2009, 19(3): 438-447.

[125] Jin Z, Yao J, Kittrell C, et al. Large-scale growth and characterizations of nitrogen-doped monolayer graphene sheets [J]. ACS Nano, 2011, 5(5): 4112-4117.

[126] Hulicova-Jurcakova D, Puziy A M, Poddubnaya O I, et al. Highly stable performance of supercapacitors from phosphorus-enriched carbons [J]. Journal of the American Chemical Society, 2009, 131(14): 5026-5027.

[127] Guo H L, Gao Q M. Boron and nitrogen co-doped porous carbon and its enhanced properties as supercapacitor [J]. Journal of Power Sources, 2009, 186(2): 551-556.

[128] Hulicova-Jurcakova D, Kodama M, Shiraishi S, et al. Nitrogen-enriched nonporous carbon electrodes with extraordinary supercapacitance [J]. Advanced Functional Materials, 2009, 19(11): 1800-1809.

[129] Chen L F, Zhang X D, Liang H W, et al. Synthesis of nitrogen-doped porous carbon nanofibers as an efficient electrode material for supercapacitors [J]. ACS Nano, 2012, 6(8): 7092-7102.

[130] Kim N D, Kim W, Joo J B, et al. Electrochemical capacitor performance

of N-doped mesoporous carbons prepared by ammoxidation [J]. Journal of Power Sources, 2008, 180(1): 671-675.

[131] Hulicova D, Yamashita J, Soneda Y, et al. Supercapacitors prepared from melamine-based carbon [J]. Chemistry of Materials, 2005, 17(5): 1241-1247.

[132] Béguin F, Szostak K, Lota G, et al. A self-supporting electrode for supercapacitors prepared by one-step pyrolysis of carbon nanotube/polyacrylonitrile blends [J]. Advanced Materials, 2005, 17(19): 2380-2384.

[133] Zhong M, Kim E K, Mcgann J P, et al. Electrochemically active nitrogen-enriched nanocarbons with well-defined morphology synthesized by pyrolysis of self-assembled block copolymer [J]. Journal of the American Chemical Society, 2012, 134(36): 14846-14857.

[134] Hu B, Wang K, Wu L H, et al. Engineering carbon materials from the hydrothermal carbonization Process of Biomass [J]. Advanced Materials, 2010, 22(7): 813-828.

[135] Iamchaturapatr J, Yi S W, Rhee J S. Nutrient removals by 21 aquatic plants for vertical free surface-flow (VFS) constructed wetland [J]. Ecological Engineering, 2007, 29(3): 287-293.

[136] Liu W J, Zeng F X, Jiang H, et al. Total recovery of nitrogen and phosphorus from three wetland plants by fast pyrolysis technology [J]. Bioresource Technology, 2011, 102(3): 3471-3479.

[137] Huber G W, Iborra S, Corma A. Synthesis of transportation fuels from biomass: chemistry, catalysts, and engineering [J]. Chemical Reviews, 2006, 106(9): 4044-4098.

[138] Armandi M, Bonelli B, Geobaldo F, et al. Nanoporous carbon materials obtained by sucrose carbonization in the presence of KOH [J]. Microporous and Mesoporous Materials, 2010, 132(3): 414-420.

[139] Wang J, Kaskel S. KOH activation of carbon-based materials for energy storage [J]. Journal of Materials Chemistry, 2012, 22(45): 23710-23725.

[140] Li Z, Zhang L, Amirkhiz B S, et al. Carbonized chicken eggshell membranes with 3D architectures as high-performance electrode materials for supercapacitors [J]. Advanced Energy Materials, 2012, 2(4): 431-437.

[141] Pollak E, Salitra G, Soffer A, et al. On the reaction of oxygen with

nitrogen-containing and nitrogen-free carbons [J]. Carbon, 2006, 44(15): 3302-3307.

[142] Paraknowitsch J P, Thomas A. Doping carbons beyond nitrogen: An overview of advanced heteroatom doped carbons with boron, sulphur and phosphorus for energy applications [J]. Energy & Environmental Science, 2013, 6(10): 2839-2855.

[143] Tao Y, Kanoh H, Abrams L, et al. Mesopore-modified zeolites: Preparation, characterization, and applications [J]. Chemical Reviews, 2006, 106(3): 896-910.

[144] Rouquerol F, Sing K. Adsorption by powders and porous solids [M]. London: Academic Press, 1999.

[145] Qie L, Chen W, Xu H, et al. Synthesis of functionalized 3D hierarchical porous carbon for high-performance supercapacitors [J]. Energy & Environmental Science, 2013, 6(8): 2497-2504.

[146] Zhao Q, Wagner H D. Raman spectroscopy of carbon-nanotube-based composites [J]. Philosophical Transactions of the Royal Society of London Series A: Mathematical, Physical and Engineering Sciences, 2004, 362(1824): 2407-2424.

[147] Ferrari A C, Robertson J. Raman spectroscopy of amorphous, nanostructured, diamond: Like carbon, and nanodiamond [J]. Philosophical Transactions of the Royal Society of London Series A: Mathematical, Physical and Engineering Sciences, 2004, 362(1824): 2477-2512.

[148] Yan J, Fan Z, Sun W, et al. Advanced asymmetric supercapacitors based on Ni(OH)$_2$/graphene and porous graphene electrodes with high energy density [J]. Advanced Functional Materials, 2012, 22(12): 2632-2641.

[149] Wang B, Chen J S, Wang Z, et al. Green synthesis of NiO nanobelts with exceptional pseudo-capacitive properties [J]. Advanced Energy Materials, 2012, 2(10): 1188-1192.

[150] Raymundo-Piñero E, Leroux F, Béguin F. A high-performance carbon for supercapacitors obtained by carbonization of a seaweed biopolymer [J]. Advanced Materials, 2006, 18(14): 1877-1882.

[151] Raymundo-Piñero E, Cadek M, Beguin F. Tuning carbon materials for supercapacitors by direct pyrolysis of seaweeds [J]. Advanced Functional

Materials, 2009, 19(7): 1032-1039.

[152] Wei L, Sevilla M, Fuertes A B, et al. Hydrothermal carbonization of abundant renewable natural organic chemicals for high-performance supercapacitor electrodes [J]. Advanced Energy Materials, 2011, 1(3): 356-361.

[153] Frackowiak E. Carbon materials for supercapacitor application [J]. Physical Chemistry Chemical Physics, 2007, 9(15): 1774-1785.

[154] Conway B E. Electrochemical supercapacitors [M]. New York: Kluwer Academic/Plenum Publishers, 1999.

[155] Kim Y J, Abe Y, Yanagiura T, et al. Easy preparation of nitrogen-enriched carbon materials from peptides of silk fibroins and their use to produce a high volumetric energy density in supercapacitors [J]. Carbon, 2007, 45(10): 2116-2125.

[156] Béguin F, Szostak K, Lota G, et al. A self-supporting electrode for supercapacitors prepared by one-step pyrolysis of carbon nanotube/polyacrylonitrile blends [J]. Advanced Materials, 2005, 17(19): 2380-2384.

[157] Lota G, Lota K, Frackowiak E. Nanotubes based composites rich in nitrogen for supercapacitor application [J]. Electrochemistry Communications, 2007, 9(7): 1828-1832.

[158] Frackowiak E, Lota G, Machnikowski J, et al. Optimisation of supercapacitors using carbons with controlled nanotexture and nitrogen content [J]. Electrochimica Acta, 2006, 51(11): 2209-2214.

[159] Li W R, Chen D H, Li Z, et al. Nitrogen enriched mesoporous carbon spheres obtained by a facile method and its application for electrochemical capacitor [J]. Electrochemistry Communications, 2007, 9(4): 569-573.

[160] Su F, Poh C K, Chen J S, et al. Nitrogen-containing microporous carbon nanospheres with improved capacitive properties [J]. Energy & Environmental Science, 2011, 4(3): 717-724.

[161] Wei J, Zhou D D, Sun Z K, et al. A controllable synthesis of rich nitrogen-doped ordered mesoporous carbon for CO_2 capture and supercapacitors [J]. Advanced Functional Materials, 2013, 23: 2322-2328.

[162] Chen S, Zhu J W, Wu X D, et al. Graphene oxide-MnO_2 nanocomposites for supercapacitors [J]. ACS Nano, 2010, 4(5): 2822-2830.

[163] Yan J, Khoo E, Sumboja A, et al. Facile Coating of Manganese Oxide on Tin Oxide Nanowires with High-Performance Capacitive Behavior [J]. ACS Nano, 2010, 4(7): 4247-4255.

[164] Zhu Y, Murali S, Stoller M D, et al. Carbon-based supercapacitors produced by activation of graphene [J]. Science, 2011, 332(6037): 1537-1541.

[165] Liu C, Li F, Ma L P, et al. Advanced materials for energy storage [J]. Advanced Materials, 2010, 22(8): E28-E62.

[166] Lin Y H, Wei T Y, Chien H C, et al. Manganese oxide/carbon aerogel composite: An outstanding supercapacitor electrode material [J]. Advanced Energy Materials, 2011, 1(5): 901-907.

[167] Wang L, Ye Y J, Lu X P, et al. Hierarchical nanocomposites of polyaniline nanowire arrays on reduced graphene oxide sheets for supercapacitors [J]. Sci Rep, 2014, 203: 864-872.

[168] Li X, Rong J P, Wei B Q. Electrochemical Behavior of Single-Walled Carbon Nanotube Supercapacitors under Compressive Stress [J]. ACS Nano, 2010, 4(10): 6039-6049.

[169] Ania C O, Khomenko V, Raymundo-Piñero E, et al. The large electrochemical capacitance of microporous doped carbon obtained by using a zeolite template [J]. Advanced Functional Materials, 2007, 17(11): 1828-1836.

[170] Wu C C, Li H Q, Xia Z X, et al. NiFe layered double hydroxides with unsaturated metal sites via precovered surface strategy for oxygen evolution reaction [J]. ACS Catalysis, 2020, 10(19): 11127-11135.

[171] Liu W-J, Dang L, Xu Z, et al. Electrochemical oxidation of 5-hydroxymethylfurfural with NiFe layered double hydroxide (LDH) nanosheet catalysts [J]. ACS Catalysis, 2018, 8(6): 5533-5541.

[172] Ahmed M J, Hameed B H, Hummadi E H. Review on recent progress in chitosan/chitin-carbonaceous material composites for the adsorption of water pollutants [J]. Carbohydr Polym, 2020, 247: 116690.

[173] El Knidri H, Belaabed R, Addaou A, et al. Extraction, chemical modification and characterization of chitin and chitosan [J]. Int J Biol Macromol, 2018, 120(Pt A): 1181-1189.

[174] Yadaei H, Beyki M H, Shemirani F, et al. Ferrofluid mediated chitosan-

mesoporous carbon nanohybrid for green adsorption/preconcentration of toxic Cd(Ⅱ): Modeling, kinetic and isotherm study [J]. Reactive and Functional Polymers, 2018, 122: 85-97.

[175] González J A, Villanueva M E, Piehl L L, et al. Development of a chitin/graphene oxide hybrid composite for the removal of pollutant dyes: Adsorption and desorption study [J]. Chemical Engineering Journal, 2015, 280: 41-48.

[176] Wei X, Ma K, Cheng Y, et al. Adhesive, conductive, self-healing, and antibacterial hydrogel based on chitosan-polyoxometalate complexes for Wearable Strain Sensor [J]. ACS Applied Polymer Materials, 2020, 2(7): 2541-2549.

[177] Shigehiko Y, Sato H. Some physical properties of glassy carbon [J]. Nature, 1962, 193: 261-262.

[178] Zhang W L, Cao Z, Wang W X, et al. A site-selective doping strategy of carbon anodes with remarkable K-ion storage capacity [J]. Angew Chem Int Ed Engl, 2020, 59(11): 4448-4455.

[179] Galkin K I, Krivodaeva E A, Romashov L V, et al. Critical influence of 5-hydroxymethylfurfural aging and decomposition on the utility of biomass conversion in organic synthesis [J]. Angew Chem Int Ed Engl, 2016, 55(29): 8338-8342.

[180] Hayashi E, Yamaguchi Y, Kamata K, et al. Effect of MnO_2 crystal structure on aerobic oxidation of 5-hydroxymethylfurfural to 2,5-furandicarboxylic acid [J]. J Am Chem Soc, 2019, 141(2): 890-900.

[181] 陆贻超, 刘志春, 张亚杰. 2,5-呋喃二甲酸的结晶和晶体结构解析 [J]. 高校化学工程学报, 2020, 34(3): 688-696.

[182] Oturan M A, Aaron J-J. Advanced oxidation processes in water/wastewater treatment: principles and applications: A review [J]. Critical Reviews in Environmental Science and Technology, 2014, 44(23): 2577-2641.

[183] Miklos D B, Remy C, Jekel M, et al. Evaluation of advanced oxidation processes for water and wastewater treatment: A critical review [J]. Water Research, 2018, 139: 118-131.

[184] Boczkaj G, Fernandes A. Wastewater treatment by means of advanced oxidation processes at basic pH conditions: A review [J]. Chemical Engineering Journal, 2017, 320: 608-633.

[185] Munoz M, De Pedro Z M, Casas J A, et al. Preparation of magnetite-based catalysts and their application in heterogeneous Fenton oxidation: A review [J]. Applied Catalysis B: Environmental, 2015, 176-177: 249-265.

[186] Mirzaei A, Chen Z, Haghighat F, et al. Removal of pharmaceuticals from water by homo/heterogonous Fenton-type processes: A review [J]. Chemosphere, 2017, 174: 665-688.

[187] Babuponnusami A, Muthukumar K. A review on Fenton and improvements to the Fenton process for wastewater treatment [J]. Journal of Environmental Chemical Engineering, 2014, 2(1): 557-572.

[188] Brillas E, Sirés I, Oturan M A. Electro-fenton process and related electrochemical technologies based on Fenton's reaction chemistry [J]. Chemical Reviews, 2009, 109(12): 6570-6631.

[189] Ganiyu S O, Zhou M, Martínez-Huitle C A. Heterogeneous electro-Fenton and photoelectro-Fenton processes: A critical review of fundamental principles and application for water/wastewater treatment [J]. Applied Catalysis B: Environmental, 2018, 235: 103-129.

[190] Liu K, Yu J C-C, Dong H, et al. Degradation and mineralization of carbamazepine using an electro-Fenton reaction catalyzed by magnetite nanoparticles fixed on an electrocatalytic carbon fiber textile cathode [J]. Environmental Science & Technology, 2018, 52(21): 12667-12674.

[191] Ganiyu S O, Huong Le T X, Bechelany M, et al. A hierarchical CoFe-layered double hydroxide modified carbon-felt cathode for heterogeneous electro-Fenton process [J]. Journal of Materials Chemistry A, 2017, 5(7): 3655-3666.

[192] Zhao H Y, Qian L, Guan X H, et al. Continuous bulk FeCuC aerogel with ultradispersed metal nanoparticles: an efficient 3D heterogeneous electro-Fenton cathode over a wide range of pH 3~9 [J]. Environmental Science & Technology, 2016, 50(10): 5225-5233.

[193] Zhao H Y, Wang Y J, Wang Y B, et al. Electro-Fenton oxidation of pesticides with a novel Fe_3O_4-Fe_2O_3/activated carbon aerogel cathode: high activity, wide pH range and catalytic mechanism [J]. Applied Catalysis B: Environmental, 2012, 125: 120-127.

[194] Bounab L, Iglesias O, Pazos M, et al. Effective monitoring of the electro-Fenton degradation of phenolic derivatives by differential pulse

voltammetry on multi-walled-carbon nanotubes modified screen-printed carbon electrodes [J]. Applied Catalysis B: Environmental, 2016, 180: 544-550.

[195] Gao G D, Zhang Q Y, Hao Z W, et al. Carbon nanotube membrane stack for flow-through sequential regenerative electro-Fenton [J]. Environmental Science & Technology, 2015, 49(4): 2375-2383.

[196] Li J P, Ai Z H, Zhang L Z. Design of a neutral electro-Fenton system with Fe-Fe_2O_3/ACF composite cathode for wastewater treatment [J]. Journal of Hazardous Materials, 2009, 164(1): 18-25.

[197] Deng F X, Li S X, Zhou M H, et al. A biochar modified nickel-foam cathode with iron-foam catalyst in electro-Fenton for sulfamerazine degradation [J]. Applied Catalysis B: Environmental, 2019, 256: 117796.

[198] Zhu Y P, Zhu R L, Xi Y F, et al. Strategies for enhancing the heterogeneous Fenton catalytic reactivity: A review [J]. Applied Catalysis B: Environmental, 2019, 255: 117739.

[199] Zhang C, Zhou M H, Ren G B, et al. Heterogeneous electro-Fenton using modified iron-carbon as catalyst for 2, 4-dichlorophenol degradation: Influence factors, mechanism and degradation pathway [J]. Water Research, 2015, 70: 414-424.

[200] Sun J K, Xu Q. Functional materials derived from open framework templates/precursors: Synthesis and applications [J]. Energy & Environmental Science, 2014, 7(7): 2071-2100.

[201] Ahn S H, Yu X, Manthiram A. "Wiring" Fe-Nx-embedded porous carbon framework onto 1D nanotubes for efficient oxygen reduction reaction in alkaline and acidic media [J]. Advanced Materials, 2017, 29(26): 1606534.

[202] Liu W J, Jiang H, Yu H Q. Development of biochar-based functional materials: Toward a sustainable platform carbon material [J]. Chemical Reviews, 2015, 115(22): 12251-12285.

[203] Zhang S, Dokko K, Watanabe M. Carbon materialization of ionic liquids: From solvents to materials [J]. Materials Horizons, 2015, 2(2): 168-197.

[204] Deng F, Olvera-Vargas H, Garcia-Rodriguez O, et al. Waste-wood-derived biochar cathode and its application in electro-Fenton for sulfathiazole treatment at alkaline pH with pyrophosphate electrolyte [J].

Journal of Hazardous Materials, 2019, 377: 249-258.

[205] Rouquerol F, Rouquerol J, Sing K S W. 2-thermodynamics of adsorption at the gas/solid interface[M]//Rouquerol F, Rouquerol J, Sing K S W, et al. Adsorption by Powders and Porous Solids. 2nd ed. Oxford: Academic Press, 2014.

[206] Liu W J, Li W W, Jiang H, et al. Fates of chemical elements in biomass during its pyrolysis [J]. Chemical Reviews, 2017, 117(9): 6367-6398.

[207] Liu W J, Tian K, He Y R, et al. High-yield harvest of nanofibers/mesoporous carbon composite by pyrolysis of waste biomass and its application for high durability electrochemical energy storage [J]. Environmental Science & Technology, 2014, 48(23): 13951-13959.

[208] Liu W J, Tian K, Jiang H, et al. Harvest of Cu NP anchored magnetic carbon materials from Fe/Cu preloaded biomass: Their pyrolysis, characterization, and catalytic activity on aqueous reduction of 4-nitrophenol [J]. Green Chemistry, 2014, 16(9): 4198-4205.

[209] Xiao J W, Xu Y, Xia Y, et al. Ultra-small Fe_2N nanocrystals embedded into mesoporous nitrogen-doped graphitic carbon spheres as a highly active, stable, and methanol-tolerant electrocatalyst for the oxygen reduction reaction [J]. Nano Energy, 2016, 24: 121-129.

[210] Lin L, Zhu Q, Xu A W. Noble-metal-free Fe-N/C catalyst for highly efficient oxygen reduction reaction under both alkaline and acidic conditions [J]. Journal of the American Chemical Society, 2014, 136(31): 11027-11033.

[211] Chung H T, Won J H, Zelenay P. Active and stable carbon nanotube/nanoparticle composite electrocatalyst for oxygen reduction [J]. Nature Communications, 2013, 4(1): 1922.

[212] Pan Z, Wang K, Wang Y, et al. In-situ electrosynthesis of hydrogen peroxide and wastewater treatment application: A novel strategy for graphite felt activation [J]. Applied Catalysis B: Environmental, 2018, 237: 392-400.

[213] Jiang S J, Li Z, Wang H Y, et al. Tuning nondoped carbon nanotubes to an efficient metal-free electrocatalyst for oxygen reduction reaction by localizing the orbital of the nanotubes with topological defects [J]. Nanoscale, 2014, 6(23): 14262-14269.

[214] Xu P, Xu H, Zheng D Y. The efficiency and mechanism in a novel electro-Fenton process assisted by anodic photocatalysis on advanced treatment of coal gasification wastewater [J]. Chemical Engineering Journal, 2019, 361: 968-974.

[215] Zhao H Y, Qian L, Chen Y, et al. Selective catalytic two-electron O_2 reduction for onsite efficient oxidation reaction in heterogeneous electro-Fenton process [J]. Chemical Engineering Journal, 2018, 332: 486-498.

[216] Mi X Y, Han J J, Sun Y, et al. Enhanced catalytic degradation by using RGO-Ce/WO_3 nanosheets modified CF as electro-Fenton cathode: Influence factors, reaction mechanism and pathways [J]. Journal of Hazardous Materials, 2019, 367: 365-374.